MIX
Papier aus verantwortungsvollen Quellen
Paper from responsible sources
FSC® C105338

Maya Yadav

Asbestos Exposure in India

Anchor Academic
Publishing

Yadav, Maya: Asbestos Exposure in India, Hamburg, Anchor Academic Publishing 2016

Buch-ISBN: 978-3-95489-453-6
PDF-eBook-ISBN: 978-3-95489-488-8
Druck/Herstellung: Anchor Academic Publishing, Hamburg, 2016
Covermotiv: © Maya Yadav

Bibliografische Information der Deutschen Nationalbibliothek:
Die Deutsche Nationalbibliothek verzeichnet diese Publikation in der Deutschen Nationalbibliografie; detaillierte bibliografische Daten sind im Internet über http://dnb.d-nb.de abrufbar.

Bibliographical Information of the German National Library:
The German National Library lists this publication in the German National Bibliography. Detailed bibliographic data can be found at: http://dnb.d-nb.de

All rights reserved. This publication may not be reproduced, stored in a retrieval system or transmitted, in any form or by any means, electronic, mechanical, photocopying, recording or otherwise, without the prior permission of the publishers.

Das Werk einschließlich aller seiner Teile ist urheberrechtlich geschützt. Jede Verwertung außerhalb der Grenzen des Urheberrechtsgesetzes ist ohne Zustimmung des Verlages unzulässig und strafbar. Dies gilt insbesondere für Vervielfältigungen, Übersetzungen, Mikroverfilmungen und die Einspeicherung und Bearbeitung in elektronischen Systemen.

Die Wiedergabe von Gebrauchsnamen, Handelsnamen, Warenbezeichnungen usw. in diesem Werk berechtigt auch ohne besondere Kennzeichnung nicht zu der Annahme, dass solche Namen im Sinne der Warenzeichen- und Markenschutz-Gesetzgebung als frei zu betrachten wären und daher von jedermann benutzt werden dürften.

Die Informationen in diesem Werk wurden mit Sorgfalt erarbeitet. Dennoch können Fehler nicht vollständig ausgeschlossen werden und die Diplomica Verlag GmbH, die Autoren oder Übersetzer übernehmen keine juristische Verantwortung oder irgendeine Haftung für evtl. verbliebene fehlerhafte Angaben und deren Folgen.

Alle Rechte vorbehalten

© Anchor Academic Publishing, Imprint der Diplomica Verlag GmbH
Hermannstal 119k, 22119 Hamburg
http://www.diplomica-verlag.de, Hamburg 2016
Printed in Germany

ACKNOWLEDGEMENT

I would like to use this opportunity to express my gratitude to everyone who supported me throughout my research work.

Firstly, I would like to thank my supervisor **Dr. rer. nat. Jörn H. Kruhl** for his endeavor approach and outstanding supervision during my research. I could not have imagined having a better supervisor. I would like to express my deep gratitude to my supervisor for giving me this opportunity to work on such a challenging topic. His guidance and constructive feedback helped me to continuously improve my research work.

I would also like to thank **Dr. Md. Sakatwat Hossain**, for his consistent support in quantifying my laboratory analyses.

Most importantly, I would also like to thank **Ms. Alexandra Caterbow**, Senior Policy Advisor from WECF (Women in Europe for a Common Future) for her invaluable cooperation and direction in getting recent information.

I would particularly like to thank **Mr. Mohit Gupta, Mr. Krishnendu Mukherjee, Mr. Sanjiv Pandita, Ms. Laurie Kazan-Allen, Ms. Madhumita Dutta, Dr. Tushar Joshi** and **Mr. Gopal Krishna** for providing me recent data and information about India.

Finally my heartfelt thanks to my parents for their blessings and insurmountable help which I have received over the years, without which this would just not have been possible.

ABSTRACT

Despite the fact that ban on mining in India being placed for more than 20 years, mines in the private sector of India are still in operation and asbestos continues to be used in large quantities. The primary purpose of this study is to provide a range of background information about asbestos and to assess and quantify asbestos from asbestos-containing materials. Data for this study were collected and obtained from International Labor Organization (ILO), World Health Organization (WHO) and Non-Governmental Organizations (NGOs) like the Ban Asbestos Network of India (BANI), International Ban Asbestos Secretariat (IBAS), Women in Europe for a Common Future (WECF), Asian Network for the Rights of Occupational Accident Victims (ANROAV), on asbestos exposure and prevalence of asbestos in India; in addition SEM analysis was done to quantify the asbestos cement sample obtained from Abfallwirtschaftsbetrieb München (AWM). Asbestos is now banned in more than 50 countries and safer products have replaced many materials that once were made with asbestos. Nonetheless, a large number of countries still use, import, and export asbestos and asbestos-containing materials. It was concluded that all forms of asbestos pose grave to human health. All are proven human carcinogens. There is no continued justification for the use of asbestos. Its production and use should be banned worldwide. A global ban on asbestos is needed.

Keywords: Asbestos, Carcinogenic, Fiber, Asbestos-containing materials, Alternatives, Ban, SEM.

TABLE OF CONTENTS

ACKNOWLEDGEMENT 7
ABSTRACT 9
LIST OF FIGURES 13
LIST OF TABLES 13
LIST OF APPENDICES 13
LIST OF ABBREVIATIONS 14
1   INTRODUCTION 15
   1.1   Research Background 15
   1.2   Research Objective 16
   1.3   Structure of the Report 16
2   LITERATURE REVIEW 17
   2.1   Background 17
   2.2   Types of Asbestos 18
      2.2.1   Serpentine Group 18
      2.2.2   Amphibole Group 19
   2.3   Serpentine and Amphibole Crystal structure and shape 21
   2.4   Asbestos deposits in India 22
   2.5   Consumption of asbestos fibers in India 25
   2.6   Import and export of asbestos in India 25
3   ASBESTOS CONTAINING MATERIALS 27
   3.1   Introduction 27
   3.2   Risk and precaution of asbestos exposure 29
   3.3   Approved Laboratories in India 30
   3.4   Methods of Identifying Asbestos fibers 32
      3.4.1   SEM (Scanning Electron Microscopy) 32
      3.4.2   TEM (Transmission Electron Microscopy) 33
   3.5   Laboratory analyses of asbestos cement sample 34
      3.5.1   Results of SEM Analysis 35

4 HEALTH IMPACTS DUE TO ASBESTOS EXPOSURE IN INDIA 48
4.1 Toxicity of various types of Asbestos 48
4.2 Asbestos Related Disease (ARD) 48
4.2.1 Asbestosis 49
4.2.2 Malignant Mesothelioma 49
4.2.3 Asbestos Related Lung Cancer (Bronchial carcinoma) 50
4.3 Occupational exposure in India 51

5 INDIAN REGULATION ON ASSESSMENT AND MANAGEMENT OF ASBESTOS EXPOSURE 52
5.1 Waste Handling 52
5.1.1 Waste Avoidance 53
5.1.2 Waste Collection 54
5.1.3 Identification and Isolation of waste 55
5.1.4 Transportation of Asbestos waste 55
5.1.5 Disposal of asbestos waste 55
5.1.6 Personal protection and hygiene 56
5.1.7 Supervision 56

6 SAFER ALTERNATIVES OF ASBESTOS 57
6.1 List of asbestos alternatives 58

7 RESULTS AND DISCUSSION 61

8 CONCLUSION 63

9 REFERENCES 65

10 APPENDICES 72

LIST OF FIGURES

Figure 1:	Types of Asbestos	18
Figure 2:	Amphiboles crystals structure	21
Figure 3:	Serpentine crystals structure	21
Figure 4:	Asbestos mines in India	24
Figure 5:	Global asbestos fiber consumption, 2012	25
Figure 6:	Asbestos roofing in India	29
Figure 7:	Binocular photographs of asbestos cement sample	34
Figure 8(a-d):	SEM Images of asbestos cement sample	35
Figure 9:	SEM image showing single asbestos fibers	38
Figure 10:	Energy dispersive x-ray spectrum showing elemental peaks	38
Figure 11:	SEM image showing bundle of asbestos fibers	40
Figure 12:	Energy dispersive x-ray spectrum showing elemental peaks	40
Figure 13:	Histogram showing Fiber length in μm	42
Figure 14:	Histogram showing Fiber width in μm	42
Figure 15:	Analysis of asbestos fiber to determine its fiber complexity	46
Figure 16:	Intercept ellipse with axial ratio	47
Figure 17:	OSHA compliant respirators and personal protective equipment	53

LIST OF TABLES

Table 1:	Asbestos Production (Quantity), 2011-12 and 2012-13 (By Sectors)	22
Table 2:	Results of EDXA Spectrum	39
Table 3:	Results of EDXA Spectrum	41
Table 4:	List of Asbestos Alternatives	60

LIST OF APPENDICES

Appendix 1:	Fiber length quantification from Single Fibers	72
Appendix 2:	Fiber width quantification from Single Fibers	73
Appendix 3:	SEM Image of MY_1-2-3_SE and MY_1-1-4_SE	73
Appendix 4:	Standards specified under Factory Act and Mines Act	74
Appendix 5:	Properties of various types of Asbestos	76

LIST OF ABBREVIATIONS

ANROEV: Asian Network for the Rights of Occupational and Environmental Victims
ATSDR: Agency for Toxic Substances and Disease Registry
AWM: Abfallwirtschaftsbetrieb München
BIS: Bureau of Indian Standards
DTE: Down to Earth
EDXA: Energy Dispersive X-ray Analysis
EIA: Environmental Impact Assessment
GIA: Gemological Institute of America
IARC: International Agency for Research on Cancer
IBAS: International Ban Asbestos Secretariat
ILO: International Labor Organization
MCDM: Modified Cantor-Dust Method
NABL: National Accreditation Board for Testing and Calibration Laboratories
NIOSH: National Institute for Occupational Safety and Health
OEHNI: Occupational and Environmental Health Network of India
OSHA: Occupational Safety and Health Administration
PCM: Phased Contrast Microscope
PIC: Prior Informed Consent
PLM: Polarized Light Microscopy
PPE: Personal Protective Equipment
PVA: Poly- Vinyl Acetate
PVC: Poly-Vinyl Chloride
SAED: Selected Area Electron Diffraction
SEM: Scanning Electron Microscopy
TEM: Transmission Electron Microscopy
US EPA: United States Environmental Protection Agency
WHO: World Health Organization
XRD: X-Ray Diffraction

1 INTRODUCTION

1.1 Research Background

Asbestos is banned in most industrialized countries. Yet it is still the biggest occupational killer worldwide: the World Health Organization (WHO) estimates asbestos causes more than 107,000 deaths globally every year through occupational exposure alone [WHO, 2014a].

More than 90 per cent of global asbestos production and trade are associated with chrysotile asbestos (or white asbestos) - a serpentine mineral of tabular silicates subclass [WHO, 2014b; Speranskaya et al., 2008]. There is a scientific consensus based on conclusive proof that all types of asbestos are hazardous to human health. The International Agency for Cancer Research (IARC) classifies asbestos as a proven human carcinogen. No amount of asbestos exposure is safe for human health [IARC, 2012]. Consequently, the use of all forms of asbestos has been banned in most of the developed countries. Unfortunately, India has failed to impose such ban. On contrary, India has greatly increased asbestos use in recent years. The case of asbestos use in India is clear example of a "Crime against Humanity" where the government and the asbestos industry with full knowledge of the harmful effects of asbestos, are allowing millions of people to be exposed to this deadly toxic substance [Allen, 2007].

To this day, it's still confidential to the workers about the health vulnerabilities of asbestos and are faced with scandals in government efforts to deal with public health asbestos problems. Asbestos is sold without statutory warning symbol in the market of India and are not penalized and in majority cases the workers do not wear the protection wear [DTE, 2000].

There is a very little knowledge in India and many other countries about asbestos and its health effects. One might think that since this topic is so well scientifically explored in the EU, there is a spillover effect to other countries. However, the current Indian government makes no attempt to learn the lessons from for example: the EU. To spark regulating efforts in India, and other countries, it is necessary to show data and research from national level. Therefore more research about the national situation is needed, combined with already existing studies from other parts of the world.

Throughout the last few years, there has not been any detailed analytical study over asbestos exposure in India. Motivated by these findings, the focus of this study is to further explore within the context of asbestos exposure and its fiber morphology- how the presence of asbestos fiber influences the human health, what retailers needs to be aware of and consider regarding asbestos exposure and should "reveal" the facts to the workers as well as citizens both literally and visually.

1.2 Research Objective

Many studies in India have been focused on the importance of banning asbestos production, and asbestos-related diseases, while offering no explicit information on asbestos fibers carcinogenic potentiality and asbestos waste management. So far, no comprehensive study of the asbestos and asbestos waste management has been performed in India. This study, therefore, aims to focus on the fiber carcinogenetic to human health, difficulties in managing asbestos waste, and to propose various asbestos alternative "safer" products specified by number of organizations.

Therefore one can better understand and analyze the various attributes of the asbestos exposure, and can be aware of and avoid the future use of asbestos and manage damage asbestos-containing materials with necessary precaution.

1.3 Structure of the Report

This report has been divided into 8 chapters and organized as follows. The chapter 1 provides an introduction to the research study briefly outlining the research background and objectives of the study and finally summarizes the overall structure of the report. Literature on asbestos along with recent data on production and consumption of asbestos in India has been reviewed in chapter 2; Chapter 3 describes the risk and precaution of asbestos exposure, different methods of identifying asbestos fibers and laboratory analyses of asbestos cement sample; chapter 4 discusses about the health impacts due to asbestos exposure in both occupational and non-occupation environment; Chapter 5 is a framework incorporating points that can be used to heap address the Indian regulation for asbestos waste management; chapter 6 contains the safer alternatives of asbestos; chapter 7 contains main results and discussion and finally chapter 8 presents the conclusion of the main findings and puts the research into a wider context. It describes the contribution of this study, recommends fields for further research.

2 LITERATURE REVIEW

2.1 Background

Asbestos once has been considered as a miracle material due to its excellent heat and fire resistant qualities, in addition, has a history that dates back to the prehistoric Greek Island of Ewoia- which is assumed to be the site of the first asbestos mine. As a matter of fact, the word "asbestos" comes from a Greek word meaning "inextinguishable" [Hylton et al., 2008].

According to World Health Organization [WHO, 2014], asbestos is defined as-

The term "Asbestos" is a common name for a group of naturally occurring fibrous minerals with current or historical commercial usefulness due to their extraordinary tensile strength, poor heat conduction, and relative resistance to chemical attack. For these reasons, asbestos is used for insulation in buildings and as an ingredient in a number of products, such as roofing shingles, water supply lines and fire blankets, as well as clutches and brake linings, gaskets and pads for automobiles.

The continuous use of enormous quantities of asbestos around the world remains as an important commodity in the global trade [Frank and Joshi, 2014]. It is well conveyed that the exposure to asbestos causes negative health effects on people who are occupationally exposed, work or live in asbestos-containing buildings, and or living close to the asbestos source. Many cases of secondary exposure are known, e.g. wives washing the working cloths of their husbands. In India only, about 2 to 3 million active workers are suffering from exposure to asbestos and other dusts or fibers [Joshi and Gupta, 2004]. The number of suspected people who were exposed to asbestos in India was 10 million till 2002 [Allred, 2003]. According to the International Labor Organization (ILO), a million people will die due to asbestos-related cancers by 2020 [ILO, 1999; Joshi and Gupta, 2004]. Asbestos-related diseases have a very long latency period of 20 to 40 years. In Germany, for example, where asbestos of all types were banned in 1993, the peak of ARD were recognized since 1980 [IBAS, 2015; Federal Institute for Occupational Safety and Health, 2014]. In India, although it has been banned since 1993 [Sreedhar and Alag, 2014] there is no such restriction on the use of asbestos. On contrary, there is an increasing consumption rate of asbestos in India [Joshi and Gupta, 2004; Ramanathan and Subramanian, 2001]. Asbestos exposure is clearly related to serious morbidity and early mortality [Jadhav and Roy 2012].

2.2 Types of Asbestos

Asbestos is commonly composed of fiber bundles which can be effortlessly separated into long, and thin fibers. Conclusive identification of a particular fiber type depends upon the microscopic study and detailed analysis. United States Environmental Protection Agency (US EPA) has defined six mineral types of "asbestos" including those belonging to the serpentine group and those belonging to the amphibole group. All six asbestos mineral types are known to be human carcinogens [IARC, 2012].

Asbestos minerals are mainly divided into 2 groups—amphibole and serpentine group—based on their chemistry and fiber morphology.

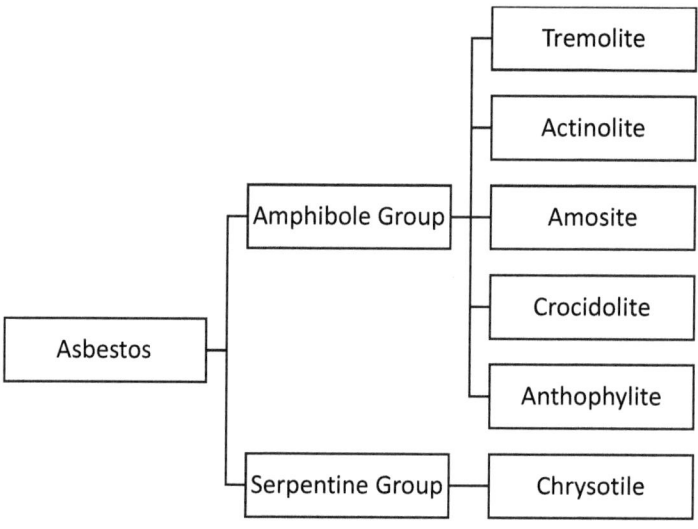

Figure 1: Types of Asbestos

2.2.1 Serpentine Group

The serpentine group is comprised solely of chrysotile asbestos. Serpentine fibers appear wavy under low magnification [Frank and Joshi, 2014].

2.2.1.1 Chrysotile [$Mg_3Si_2O_5(OH)_4$]

Chrysotile, also known as white asbestos, accounts for some 90-95 per cent of all the asbestos used worldwide [Frank and Joshi, 2014]. Chrysotile is a fibrous hydrated magnesium silicate mineral [Mg Si (OH)], which is used, in about 3,000 commercial products worldwide [Ramanathan and Subramanian, 2001]. Chrysotile, the commercial variety of asbestos is known to cause mesothelioma [Lemen, 2004]. Even though, chrysotile is the most ordinary type of asbestos, it accounts for over 95 per cent of the world production and is exploited in more than 40 countries [Ansari et al., 2007]. Asbestos cement industry is the largest user of chrysotile asbestos throughout the world and asbestos cement products are made stronger by the addition of approximately 5 to 10 per cent of chrysotile asbestos during mixing of cement [Dave and Beckett, 2005]. India consumes about 0.1 million tons of chrysotile every year, mostly imported from Canada, Brazil, Kazakhstan, Russia and South Africa [Allen, 2005]. In Bihar (India), chrysotile asbestos occurs in Singhbhum districts associated with serpentinised dunites and peridodites and is typically between 3.1 to 6.2 mm long and short between 9.4 to 15.75 mm. Likewise, Lakshmana mines in Cuddapah district in Andrapradesh chrysotile fibers are between 0.076 to 0.152 mm in length [Ban Asbestos India, 2007].

2.2.2 Amphibole Group

The amphibole group consist of amosite, crocidolite, anthophyllite, tremolite, and actinolite asbestos. All the amphibole fibers are straight and needle-like in their microscopic appearance [Frank and Joshi, 2014]

2.2.2.1 Amosite [$Fe_7Si_8O_{22}(OH)_2$]

Amosite is perceived as a grey-white vitreous fiber under a microscope, therefore it is also known as the mineral grunerite or brown asbestos and contain iron and magnesium [Mukherjee, 2012]. It is found most frequently as a fire retardant in thermal insulation products, asbestos insulating board (which contained up to 40 percent asbestos) and ceiling tiles [Wisconsin Department of Natural Resources, 2007]. Like the other forms of amphibole asbestos, it has also needle-like fibers.

2.2.2.2 Crocidolite [$Na_2Fe_3^{2+}Fe_2^{3+}Si_8O_{22}(OH)_2$]

One of the various type from amphibole group, crocidolite takes the form of blue, straight fibers and has a sodium iron magnesium silicate (Na Mg Fe Si) [Mukherjee, 2012]. Crocidolite is the fibrous form of the amphibole riebeckite which is perceived as a blue fiber under a microscope [Wisconsin Department of Natural Resources, 2007]. Therefore it is also known as "blue" asbestos. Several asbestos studies suggest crocidolite may be liable for more deaths than any other type of asbestos for the reason that its fibers are so thin, about the diameter of a strand of hair. When airborne, these fibers can be breathe in easily and stuck fast in the lining of the lungs. Once inside the body, the fibers do not break down easily [Mukherjee et al, 1996]. This can lead to potentially life-threatening lungs and abdominal conditions, including Lung cancer, mesothelioma and asbestosis.

2.2.2.3 Anthophyllite [$(Mg,Fe)_7Si_8O_{22}(OH)_2$]

Anthophyllite asbestos perceived to have high potency in the carcinogenesis of lung cancer and low potency in carcinogenesis of mesothelioma in comparison with the other types of asbestos as it is formed by the breakdown of talc in ultramafic rocks in the presence of water and carbon dioxide as a prograde metamorphic reaction [Meurman et al., 1994]. The anthophyllite was used in asbestos cement and for insulation, roofing material etc. [Feininger, 2003].

2.2.2.4 Tremolite [$Ca_2Mg_5Si_8O_{22}(OH)_2$]

A form of amphibole asbestos, which is liable for many asbestos-related diseases as it is boasts sharp needle like fibers that easily enters the respiratory system when airborne. Approximately 36,500 tonnes of tremolite asbestos are mined annually in India. It is otherwise only found as a contaminant as this material is toxic and inhaling the fibers can lead to asbestosis, lung cancer and both pleural and peritoneal mesothelioma [Ansari et al, 2005].

2.2.2.5 Actinolite [$Ca_2(Mg,Fe)_5Si_8O_{22}(OH)_2$]

Actinolite is an amphibole silicate mineral, commonly found in metamorphic rocks. Like other types of amphibole asbestos, it also consists of long, sharp fibers and its make up is very similar to that of another amphibole variety i.e. tremolite asbestos. Some forms of actinolite are used as gemstones [GIA, 1988].

Fine fibers are expected higher to be breathe in than granular fibers for the reason that they remain suspended in the air for longer time period. The fine fibers that are created when asbestos is handled can penetrate deep into the lung where they can cause disease. All six types of asbestos vary in fiber size and length. The longer and finer the fiber, the bigger the danger if breathe in. Crocidolite and amosite asbestos which belongs to the amphibole group have been identified as most dangerous [Van der Perk, 2007]. Their long needle-like fibers can easily get through physically to the respiratory tract or breathing pipe and lung and in addition are most likely to cause serious health problems. Chrysotile asbestos which belongs to the serpentine group is thicker due to the curly nature of the fibres, and unable to penetrate as far. The visible fibers are themselves each composed of millions of microscopic "fibrils" that can be released by abrasion and other processes [Gee and Greenberg, 2002]. However, there is scientifically consensus that chrysotile asbestos is carcinogenic, and can cause cancer to the lungs, thyroid system and even the ovary [IARC, 2012; WHO, 2014].

2.3 Serpentine and Amphibole Crystal structure and shape

Toxicity highly depends on shape and size of the crystals. The crystal structures of the two main asbestos forming minerals, the serpentine and amphibole are surprisingly very different.

The amphiboles are "chain silicates" in which S_iO_4 tetrahedron are linked to form bands four tetrahedral wide and very ling. These bands runs parallel to asbestos fiber axis. Amphibole crystals are longer relative to their width [Gravatt et al., 1978].

Figure 2: Amphiboles crystals structure

The serpentine minerals are "layered silicates" in which S_iO_4 tetrahedron are linked to thin sheets of great lateral extent. The tetrahedral all point in the same direction. When the length is extremely long compared with the width, the crystals are called asbestiform or fibrous [Virta, 2001; Gravatt et al., 1978].

Figure 3: Serpentine crystals structure

Unlike serpentine, amphibole's have a gradational transition from blocky to prismatic to acicular to asbestiform. This gradational change makes it difficult to distinguish between asbestiform and non-asbestiform amphibole particles under the microscope. [Virta, 2001; National Bureau of Standards, 1979; Gravatt et al., 1978]

2.4 Asbestos deposits in India

There are large number of asbestos products manufacturing industries in India, both in small and large scale sectors. These industries are spread over in about fifteen major states. Nearly sixty units are in operation till date. India hosts nearly 673 small scale asbestos mining and milling facilities and 33 large scale asbestos manufacturing plants including 17 asbestos cement product manufacturing plants and 16 other than asbestos cement product manufacturing plants [Dave and Beckett, 2005; Ansari et al., 2007]. On the other hand, significant amount of small and unorganized companies are also located around the major urban centers [Biswas, 1999]. According to Indian Bureau of Mines Minerals Yearbook 2011, there were about 75 plants engaged in the production of asbestos products in the country and these are mainly located in Rajasthan (Ajmer, Bhilwara, Udaipur, Rajsamand), Jharkhand (Roro), Karnataka, Madhya Pradesh and Andra Pradesh (Pulivendla) [Subramanian and Madhavan, 2005].

Mineral: Asbestos	Production
All India Production	
2011-2012	276 tonnes
2012-2013	387 tonnes
Public Sector	-
Private Sector	
2011-2012	276 tonnes
2012-2013	387 tonnes
Overall increase/decrease in production in 2012-2013 over 2011-2012	40.22%

Table 1: Asbestos Production (Quantity), 2011-12 and 2012-13 (By Sectors)

According to Indian Bureau of Mines Mineral Yearbook 2013 on asbestos, the overall production of asbestos in 2012-2013 was 387 tons of different variety of asbestos which increased by about 40.22 per cent as compared to the previous year 2011-2012. All the mines are owned and operated by private companies [Indian Bureau of Mines Mineral Yearbook, 2013].

As shown in the figure 4, Rajasthan alone contributes nearly 62 per cent of the total Indian production of asbestos, which is industrially processed within the state itself [Indian Bureau of Mines Minerals Yearbook, 2013]. Most of the Indian asbestos deposits belong to the tremolite-actinolite variety. It occurs in tremolite-actinolite schists, amphibolites and metamorphosed basic and ultra-basic rocks. [Ban asbestos India, 2007].

Mining of asbestos is banned in India through a series of orders. In June 1986 a ban on expansion of area of existing asbestos mines was placed in the country. The letter stated - "Asbestos mining has deleterious effects on the health of the workers and exposes them to diseases like Silicosis and Pneumoconiosis etc, no expansion in the mining of Asbestos should henceforth be permitted". In March 1989 the ban was further extended to mining of those minerals as well where asbestos as contamination was found in substantial quantities and finally since 1993 all mining of asbestos has been banned [Sreedhar and Alag, 2014] Leases of all operational mines expired by 2005. 3 mines in Andhra Pradesh continue operations and report production. No any action has been taken by the government authorities to enforce the regulations, as well as no health and safety information was made available for the workers [Gupta, 2015].

Figure 4: Asbestos mines in India (*www.mapsofindia.com*)

2.5 Consumption of asbestos fibers in India

India imports nearly 100,000 metric tons of asbestos per year, and small-scale asbestos (chrysotile and tremolite) mining and milling contributes to nearly 5-10 per cent of the total national usage [Dave and Beckett, 2005].

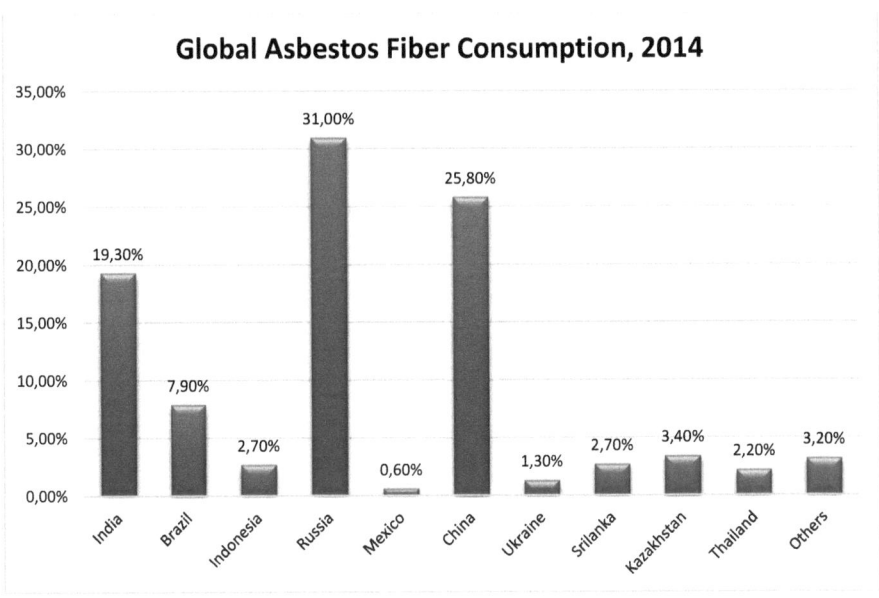

Figure 5: Global asbestos fiber consumption, 2012 [International Ban Asbestos Secretariat, 2014]

India has been a major consumer of asbestos after Russia and China [International Ban Asbestos Secretariat, 2014]. As compared with global asbestos production, India produces minimal amount of asbestos [Allen, 2007].

2.6 Import and export of asbestos in India

According to the recent United Nations Commodity Trade Statistics Database, India has become a major importer with exponential growth in manufacture of asbestos cement and pipes [Allen, 2007; United Nations Commodity Trade Statistics Database, 2014] and is primarily imported from Russia, Brazil, Canada and Kazakhstan. While as per the prevailing foreign trade policy, asbestos under heading 2524 can be imported freely with the exception of amosite which is restricted [Sreedhar and Alag, 2013-2014]. There is no restriction imposed on exports of asbestos in the amended foreign trade policy, 2009-2014 [United Nations Commodity Trade Statistics Database, 2014]. Additionally, import duties by the federal government on refractory raw minerals and mineral products have been decreased by 15 per cent in 2004 annual government budget [Subramanian and Madhavan, 2005]. On the other hand, the imports of amosite, crocidolite, tremolite, actinolite, and anthophyllite are confined in terms of Interim Prior Informed Consent (PIC) Procedure of Rotterdam Convention for Hazardous Chemicals and Pesticides, which means that import and export notification is required by international law.

According to Indian Bureau of Mines Mineral yearbook 2011, exports of asbestos were 252 tons in 2010-11 as compared to 559 tons in the previous year 2009-2010. Out of the total exports in 2010-11, exports of chrysotile asbestos were only 43 tonnes while those of other asbestos varieties were 209 tonnes. Exports of asbestos cement products were mainly to UAE, Saudi Arabia, Nepal (maximum export due to open border) and South Africa. [Sreedhar and Alag, 2013-2014]

India still silently facilitate asbestos manufacturing without serious consideration of health aspects of the worker and neighborhood's inhabitants.

3 ASBESTOS CONTAINING MATERIALS

3.1 Introduction

Asbestos is a naturally occurring mineral fiber which exists in three main forms: "brown" (amosite), "white" (chrysotile) and "blue" (crocidolite) asbestos. Asbestos possesses unique malleable strength, resistant to high temperature and flexibility from its fibers and for this reason it was widely used as binding agent to increase the strength of the products, than if just cement alone is used. Over 90 per cent of world asbestos use is in asbestos-cement pipe, flat sheet, and corrugated roofing sheet [Castleman and Joshi, 2007]. Other products still being manufactured with asbestos content include vehicle brake and clutch linings, asbestos textiles, Paints, Plastic, roofing, and gaskets sheet [Biswas, 1999]. Despite ban on asbestos mineral, asbestos is used in construction materials other than asbestos-cement products in many parts of the world.

In India, asbestos is very common in many building materials in the form of friable surfacing materials, thermal system insulation, non-friable flooring materials, and other applications especially in rural areas, and the widespread consumption of asbestos in construction materials has raised some apprehension about exposure in non-industrial environment [Guidotti, 2011]. The presence of asbestos in a building does not necessarily mean that the health of building occupants is threatened. As long as asbestos-containing materials are undisturbed, exposure is unlikely [Van Zandwijk et al., 2013] due to the reason that asbestos have the tendency to breakdown into a powder of microscopic size fibers. Because of their structure (size and shape), these tiny fibers remain pendulous in the air for longer periods of time and can easily get through physically inside body tissues after being breathe in or ingested and thereby becomes the cause of asbestos-related diseases. Conversely, damaged, or disturbed asbestos-containing materials leads to fibers release (exposure), and unauthorized removal or disturbance of asbestos materials cause's asbestos-related diseases in the long period of time. Only trained, certified workers should handle or remove asbestos-containing materials.

Asbestos-containing materials are manufactured in various states of India and has been widely used in various construction materials for a number of purposes including [Biswas, 1999; Environmental Impact assessment Guidance Manual, 2010]:

List of Asbestos-containing materials in India:

I. Asbestos Cement Products
   i. Asbestos-cement and accessories
   ii. Asbestos-cement pressure pipes and accessories
   iii. Asbestos roof sheets, slates and tiles
   iv. Corrugated asbestos cement sheets
   v. Asbestos cement building pipes, gutters and fittings (spigot and socket type)
   vi. Asbestos cement flat sheets
   vii. Asbestos cement building boards
   viii. Compressed asbestos fiber jointing
II. Gasket Sheets
   i. Compressed asbestos fiber gasket
   ii. Beater addition gaskets
III. Asbestos Clutch linings and Brake linings
IV. Asbestos Textiles
V. Special Application
   i. Millboards
   ii. Ferobestos
   iii. Plastics, Paints and Lubricants

Especially in India, corrugated asbestos cement or sheet roofing is a very popular material among slums and informal settlements of Indian cities as it is inelastic, durable, not too heavy, easy to cut and fix together, water-resistant, fire-resistant and most importantly: economical. It is indeed economical, but only in the sense of short term expenses. Comparatively it is more expensive than safer building materials as one has to pay huge amount to health care when suffered with asbestos-related diseases, which leads to even more poverty. In the other hand, there is also a triple burden on poor people concerning buying safer building materials as they do not have enough budget.

Figure 6: Asbestos roofing in India [Urban Poverty in India: Slums, 2012]

Generally, people in the slum area often cook their food in the open wood fire under asbestos roofing. This results in cracking of asbestos cement roof exposing asbestos, making it friable as height of a rural house, roofed with asbestos cement, is too small to have effective air circulation. In addition, on the roofs of Indian slums, where it is sawn and fixed by hand, it's difficult to avoid releasing a large number of dangerous fibers. They paint their asbestos roofing with emulsion paint which also deteriorates it rapidly resulting discoloration and mould growth. The effect of surface weathering also exposes asbestos from asbestos cement roof [Roy et al., 2013].

3.2 Risk and precaution of asbestos exposure

Working with, or exposure to, material that is friable, that could cause release of loose asbestos fibers, is considered to be at high risk as asbestos fibers accumulates in the lung and cause scarring and inflammation [World Bank Group, 2009]. However in general, people who are regularly exposed in a work place where they work directly with the asbestos-containing materials are considered to be at high risk of causing asbestos-related disease.

The occurrence of asbestos in household materials generally does not possesses necessarily any health hazard unless the material is fragmented, deteriorated or disturbed in such a way that airborne asbestos fibers are created. Maintenances must be done only by a skillfully professional expert in methods for securely handling the asbestos-containing materials. Maintenance usually includes either sealing or covering asbestos-containing material. Sealing (encapsulation) involves treating the material with a sealant that either binds the asbestos fibers together or coats the material so fibers are not released. Covering (enclosure) involves placing something over or around the material that contains asbestos to prevent release of fibers [Environmental Impact Assessment Guidance Manual, 2010; Woodson, 2012]. When handling asbestos material, one should take precautions to minimize the release of asbestos fibers. Some of the most important general points to be followed when handling asbestos-containing materials are as follows [BIS 11769 part 1, 1987; BIS 11769 (part 2, and 3), 1986; Woodson, 2012; Health and Safety Executive, 2012]:

- Follow each and every precaution while dealing with damaged asbestos material.
- Removal and dominant reparation should be carried by trained and qualified people in handling asbestos. It is greatly suggested that sampling and minor reparation should also be carried by asbestos professionals.
- Minimize cutting or breaking of the asbestos cement products.
- Systematically damp down the area containing asbestos material with water or a 1:10 polyvinyl acetate (PVA) before dealing with asbestos. Keep the asbestos damp up until it is packed for transportation.
- Don't powder, dust, sweep, or vacuum clean rubbles that contain asbestos.
- Avoid drilling holes in asbestos-containing materials.
- Avoid usage of scratchy pads or brushes on power strippers to strip polish from asbestos flooring. Never use a power stripper on a dry floor.
- Avoid tracking material that contain asbestos through the house. If any material is damaged, call an asbestos professional.
- Avoid asbestos cement products nearby the orchard or garden, where they might be fragmented.

When managing or treating asbestos-containing materials, householder's needs to follow the precautions defined above. According to Bureau of Indian Standards 11768 and 11769 above mentioned precautions are deliberated to decrease the danger to householders to a very minimal level. However, to ultimately avoid asbestos exposure, it is necessary to ban all forms of asbestos use. There is no controlled use of asbestos possible.

3.3 Approved Laboratories in India

To find out if a material contains asbestos, the only way is to test a sample in an approved analytical laboratory. For export there are some highly recognized laboratories in India listed below that test asbestos:

- **National Accreditation Board for Testing and Calibration Laboratories (NABL), India:** NABL is an independent organization under the Department of Science and Technology, Government of India, and is registered under the Societies Act 1860. NABL does the inter-disciplinary testing of products for e.g. textiles, plastics, building materials, metallurgical products etc. NABL has been established to provide Government, Industry associations and Industry in general with a scheme for third-party assessment of the quality and technical proficiency of testing and calibration laboratories. NABL operations follow to ISO/ IEC 17011: 2004. [Official webpage of National Accreditation Board for Testing and Calibration Laboratories, India]

- **The SGS global network, India**: SGS is the world's leading inspection, verification, testing and certification company. SGS global network of laboratories also provides quick turnaround on particulate identification state-of-the-art instruments such as XRD, QEMSCAN®, electron microprobe, optical microscopy and ICP. The SGS global network has a proven record in determining the composition and chemistry of particulates and thus helping to delineate potential process consequences and hazardous circumstances. It is also recognized as the global benchmark for quality and integrity. With more than 80,000 employees, they operate a global network of more than 1,650 offices and laboratories around the world. [Official webpage of SGS global network, India]

- **ELCA Laboratories, India**: ELCA Laboratories is one of India's successful independent laboratories, specializing in material testing which is accredited by NABL. Set up in 1974, the company provides Mechanical, Chemical, Metallurgical and Corrosion testing of Metals, Rubber, Plastics and Electrical Testing. ELCA has invested heavily and has achieved a high degree of proficiency as a result, it is one of the NABL accredited Lab in the country to identify the presence of asbestos. ELCA is dedicated to provide precise and swift testing of materials with prompt delivery of unambiguous testing reports at a reasonable cost and ensures client satisfaction. This objective is proficient by the execution of Quality Management Systems at all levels as per ISO / IEC 17025 - 2005 requirements. [Official Webpage of ELCA Laboratories, India]

- **SPECTRO Group of Companies, India**: India's primary analytical testing laboratory which begun in the year 1995, spreads its proficiency in various sectors and has existence in most important verticals including Analytical Testing Solutions, Testing Equipment Manufacturing and Marketing, Equipment repair and maintenance solution, Geotechnical services, Environment consultancy, Corporate Trainings etc. It also received NABL accreditation in the fields of Chemical, Mechanical, and Biological and Non-Destructive Testing and is certified by ISO 9001, 14001 and approved by BIS (Bureau of Indian Standard). SPECTRO constitutes the best asbestos consultants, asbestos management professionals and all services for asbestos surveys, asbestos testing and analysis, asbestos removal and asbestos training. SPECTRO is equipped with the latest technology for the analysis of asbestos fiber like PLM (Polarized Light Microscopy) technique and XRD (X-Ray Diffraction) technique. It is also engaged in projects for asbestos survey from various industries like Brake shoe industry, Fiber sheet industry and related sectors. [Official Webpage of SPECTRO Group of Companies, India]

3.4 Methods of Identifying Asbestos fibers

Asbestos fiber cannot be identified, either by means of our bare eyes or by solely observing a fiber under a regular microscope. Analytical techniques for the determination of concentrations of asbestos in the asbestos-containing materials include x-ray diffraction, differential thermos-analysis, infra-red spectrometry and other methods which do not provide fiber size information. The method which provide size information are optical, scanning electron and transmission electron microscopy due to the reason that fibers found in the asbestos-containing materials are often less than a micron in diameter and because there is no definite information on toxicity dependence with fiber size, any valid technique for asbestos fiber concentration determination must be able to provide reliable fiber species identification and accurate size measurements [National Bureau of Standards, 1977]. This limits the methods of examination to electron microscopy. These include following two methods:

- 3.4.1 SEM (Scanning Electron Microscopy)
- 3.4.2 TEM (Transmission Electron Microscopy)

3.4.1 SEM (Scanning Electron Microscopy)

Scanning Electron Microscope (SEM) using Energy Dispersive X-ray Analysis (EDXA) is one of the means of identifying asbestos fibers. The SEM is a measurement instrument that has the resolution necessary to detect very small fibers by means of electrons as a substitute of light to form an image. The SEM produces an image of the topography of the sample. The electron beam interacts with the atoms on the surface of the sample and information on the sample's composition can be collected [Graham, 2008]. The scanning electron microscope has many advantages over traditional microscopes. SEM counts the fibers and determines the elementary chemical composition [Van Zandwijk et al., 2013]. The preparation is less restrictive than with TEM as SEM analysis usually images fibers that are more than 0.2 µm in diameter because of contrast limitations [ATSDR, 2001]. The limit of analytical quantification is higher than for TEM. Ambiguities of asbestos fiber identification also arise from x-rays produced by adjacent or adhering particles, from uncertainties in determining the exact chemical composition of an asbestos mineral due to its chemical change in the environment or from the fact that a given mineral can exist over a wide range of compositions [National Bureau of Standards, 1977]. Additionally, the structure of a fiber in the electron microscope can give deceptive differences in the compositions.

3.4.2 TEM (Transmission Electron Microscopy)

Transmission Electron Microscope (TEM) identifies the elementary chemical composition and crystalline structure and counts the fibers. The TEM's electron beam passes through the sample and an image is displaced onto a screen, using an energy dispersive X-ray (EDX) and a computer system information about the fiber's composition can be gathered and graphed in their appropriate proportions [Graham, 2008]. With the help of the exact proportions of the composition in the fiber, along with additional analytical facts and figures, allows the microscope experts to distinguish various types of asbestos fiber. Selected Area Electron Diffraction (SAED) is another component of the TEM that allows for a differentiation of fibers and also allows the microscope experts to observe the diffraction patterns of the crystalline structures of the fiber being analyzed [Graham, 2008]. TEM is more precise and can be used at lower concentrations of asbestos as TEM can visualize fibers of all sizes, there is no restriction in line with SEM. It is possible to examine objects with a diameter of < 0.01 µm. It can also count fibers with a length > 0.5 µm [ATSDR, 2001].

SEM and TEM are both used largely to classify and quantify asbestos. Both SEM and TEM microscopes use a beam of electrons from a filament in a vacuum. The electron beam interacts with the atoms on the surface of the sample and information on the sample's composition is collected [Graham, 2008]. Both methods can detect smaller fibers than PCM (Phased Contrast Microscope) and also fiber type, but fiber counting accuracy is unacceptably poor.

This is a result of the small area that can be scanned at high magnification. Accuracy is more limited with long (>5 µm) fibers. [ATSDR, 2001]

3.5 Laboratory analyses of asbestos cement sample

For detailed study, I have collected an asbestos cement sample (13% asbestos and 87% Portland cement) from Waste Management Company Munich (Abfallwirtschaftsbetrieb München).

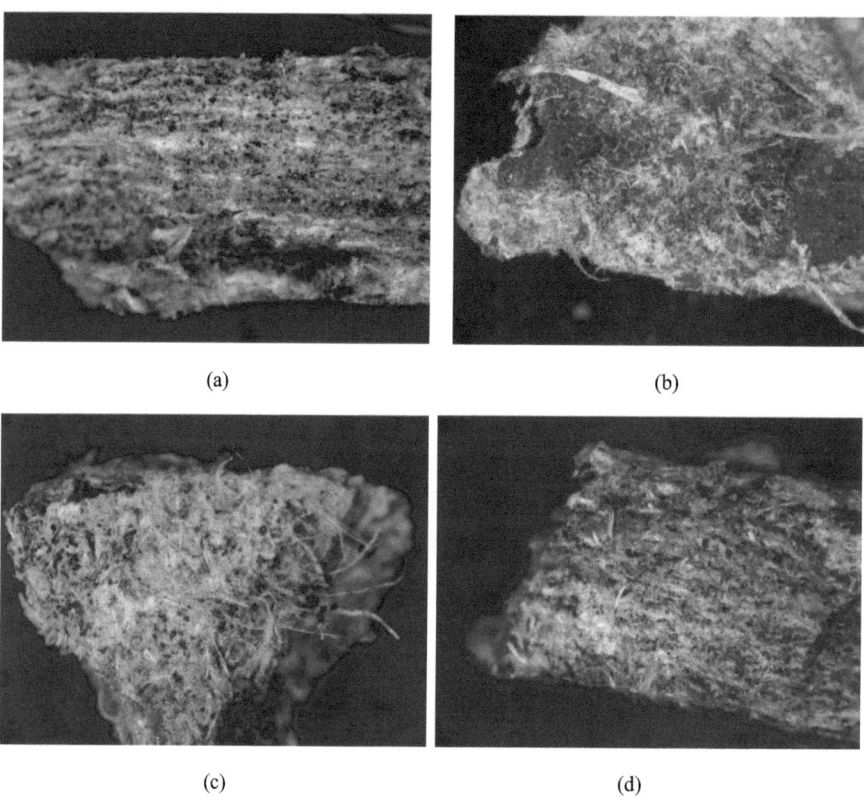

(a)   (b)

(c)   (d)

Figure 7: Binocular photographs of asbestos cement sample (First two (a and b) are parallel to the layering, the other two (c and d) are perpendicular. The preferred orientation of asbestos fibers parallel to the layering is visible and also the random orientation in the layers.
Long side of photographs is roughly 1 cm.)

Laboratory analyses were performed on the asbestos cement sample to establish a relationship between: i) morphology (shape of the crystals) ii)the asbestos cement mineralogical composition using SEM analyses and iii) High magnification structure of the individual fibers (fibrils)

iv) Energy dispersive x-ray analysis of the fiber chemistry by comparison to a standard. Also the quantification analysis of fiber length and thickness.

3.5.1 Results of SEM Analysis

3.5.1.1 Fiber Morphology and structure

The appearances of the fibers from the asbestos cement are shown in figures 8(a, b, c and d). Asbestos cement sample was randomly scanned at various magnifications to identify asbestos fibers and how asbestos fibers are mixed with construction materials. Although some fibers are seen to be straight in shape, analysis concluded that due to forcibly breaking asbestos, overlapping of grain or other matrix present in the asbestos-containing material, straight fibers turn its shape to curvy fibers. Risk of asbestos fiber release is significant.

Figure 8(a-d): SEM Images of asbestos cement sample

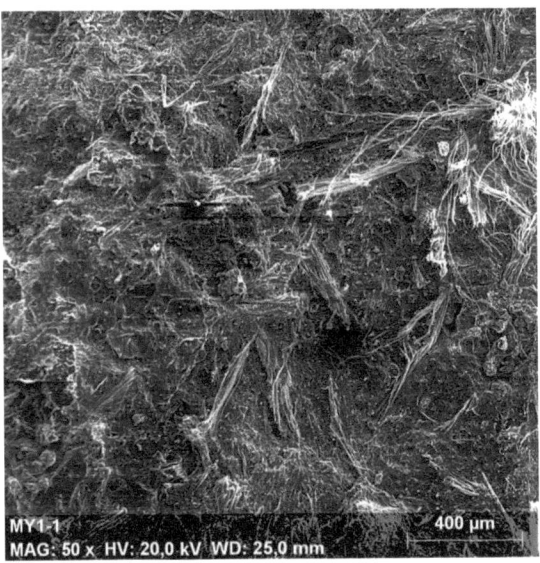

Figure 8 (a): Cement particles and asbestos fibers extracted from sample where light grey part shows asbestos fiber and dark grey part shows cement material (non-asbestos minerals).

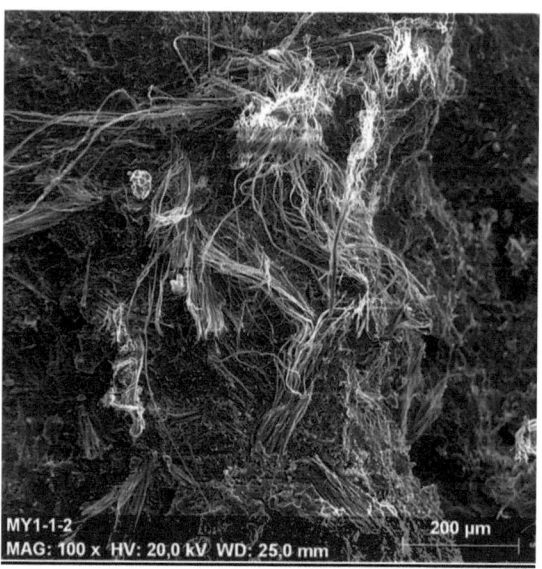

Figure 8(b): Close view of asbestos fibers where shape of the fibers are mostly irregular and a number of asbestos fibers can be seen detached from the cement material.

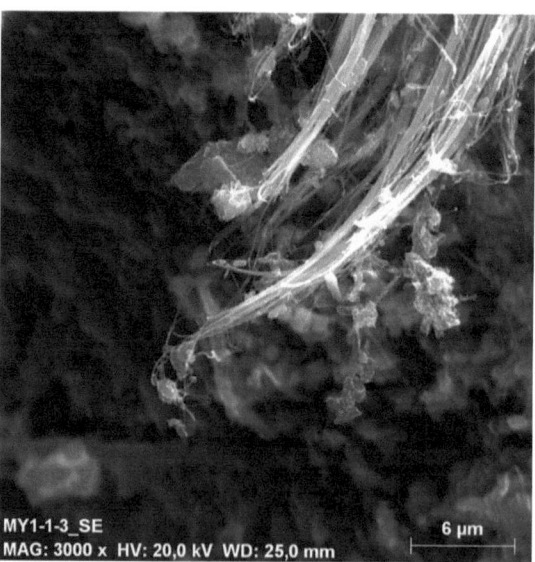

Figure 8(c): Detached asbestos fibers shows evidence of attached cement particles on the tip of the fibers bundle and are floating in the air. Fine cement particles are visible underneath.

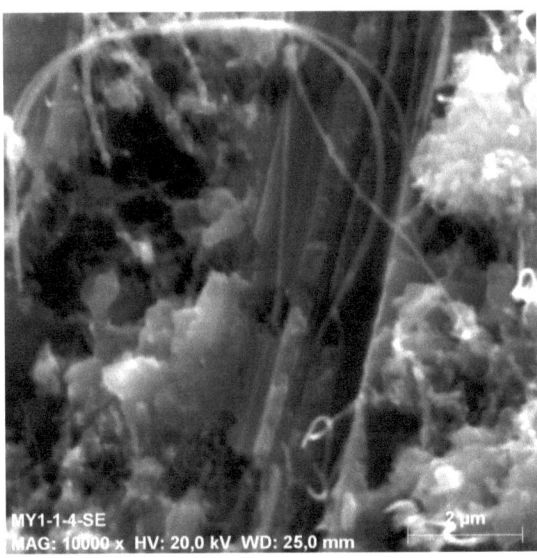

Figure 8 (d): Real shape of the individual fibers split up from fiber bundles.

3.5.1.2 Energy Dispersive X-ray Analysis (EDXA) results

EDXA is an X-ray technique used to identify the elemental composition of materials. This technique works by focusing n the electrons into a small area or probe of the fiber. The EDXA technique collects nd measures the energy of the x-rays produced and displays a graph of x-ray energy (n KeV) along the bottom axis versus frequency of occurrence. An EDXA usually shows a number of characteristic x-ray peaks, associated with the elements present. An elemental spectrum from the fiber in the sample of asbestos cement tested is shown in the figure 10.

Figure 9: SEM image showing single asbestos fibers. Area of measurement for EDX analysis is pointed as 3.

For the energy-dispersive X-ray analysis (EDXA) of area measured in SEM image (figure 9) in a scanning electron microscope, the following spectrum (figure 10) and quantitative results (table 2) were obtained.

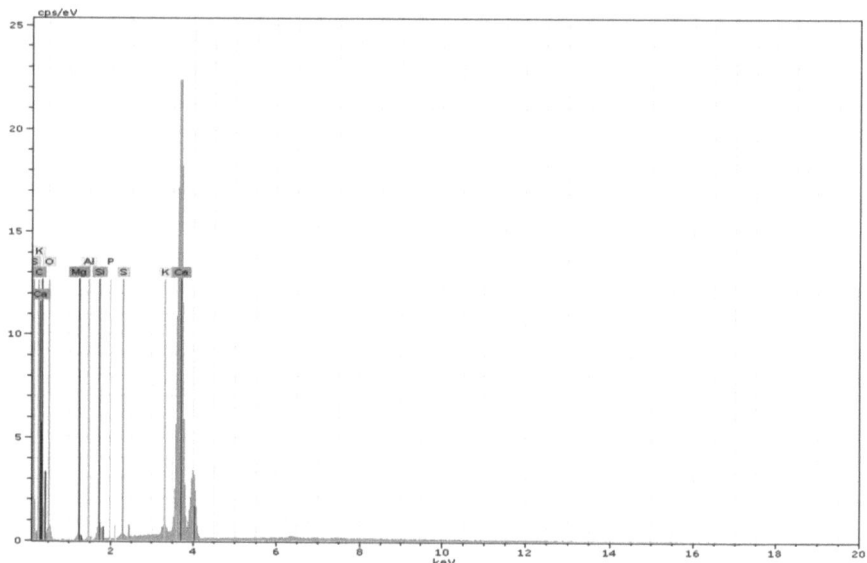

Figure 10: Energy dispersive x-ray spectrum showing elemental peaks of area measured in figure 9, where Y-axis shows the counts (number of X-rays received and processed by the detector) and the X-axis shows the energy level of those counts

Element	Series	[wt.-%]	[norm. wt.-%]	[norm. at.-%]
Carbon	K-series	3.036872271	2.947911805	6.932215325
Magnesium	K-series	0.613041032	0.595082945	0.691542069
Aluminum	K-series	0.001	0.000970707	0.00101615
Silicon	K-series	1.613382522	1.566120982	1.574997515
Phosphorus	K-series	0.001	0.000970707	0.000885178
Sulphur	K-series	0.385726655	0.374427391	0.329806756
Potassium	K-series	1.783863822	1.731608296	1.250917004
Calcium	K-series	72.43624301	70.31433554	49.55352603
Oxygen	K-series	23.14661586	22.46857163	39.66509398
	Sum:	103.0177452	100	100

Table 2: Results of EDXA Spectrum from figure 9 showing elemental composition (expressed in terms of the normalized weight in 100 per cent for each elements giving a peak above the background of continuous x-rays along with atomic %)

The X-rays radiated from the electron beam interaction with whatsoever area of the sample surface being imaged, is analyzed. The K-series basically provide the structure information from the innermost of the atom [Russ, 2013]. The K-series consists of two recognizable lines $K\alpha$ and $K\beta$ for energies above 3 keV. The approximate weights of lines in a series provide important information in identifying elements. Using standards to calibrate the detector, it is possible to carry out a quantitative analysis.

In figure 10 and table 2, the Calcium (Ca) and Potassium (K) peaks have very good peak background ratios. It can be seen that the spectrum contains small peaks of Silicon (Si) and Aluminum (Al). The Aluminum (Al) is not marked on this display but it is between Silicon and Magnesium peaks. The ability to model and accurately remove the background from the smaller intensity peaks will also affect the precision and accuracy of the measured values. Poorly defined peaks show low concentration and counting statistics [Zbyszek and Wladek, 1997].

Similarly in bundled fibers as shown in the figure 11, EDXA spectrum is calculated to analyze the difference of elemental composition between these two fibers.

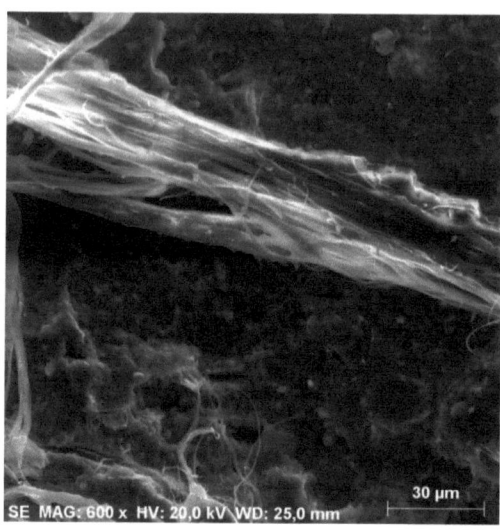

Figure 11: SEM image showing bundle of asbestos fibers split up with various thickness and are detached from the cement matrix. (For EDX analysis, area of measurement is pointed as 1 in the figure).

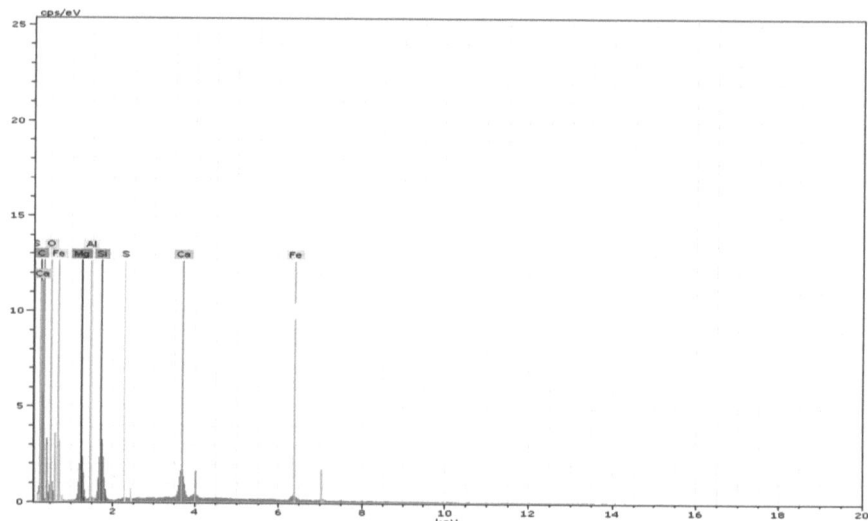

Figure 12: Energy dispersive x-ray spectrum showing elemental peaks of area measured in fig. 11, where Y-axis shows the counts (number of X-rays received and processed by the detector) and the X-axis shows the energy level of those counts.

Element	Series	[wt.-%]	[norm. wt.-%]	[norm. at.-%]
Carbon	K-series	31.6257	31.62634982	40.39776187
Magnesium	K-series	5.4501	5.450218005	3.440374066
Aluminum	K-series	0.0010	0.00100002	0.000568629
Silicon	K-series	5.3201	5.320249355	2.906278591
Sulfur	K-series	0.0463	0.046339057	0.022171231
Calcium	K-series	2.5626	2.562635601	0.980998207
Iron	K-series	0.7050	0.70505142	0.193690512
Oxygen	K-series	54.2871	54.28815672	52.05815689
	Sum:	100.0000	100	100

Table 3: Results of EDXA Spectrum from figure 11 showing elemental composition (expressed in terms of the normalized weight in 100 per cent for each elements giving a peak above the background of continuous x-rays along with atomic %.)

In figure 12 and table 3, the Calcium (Ca), Carbon (C), Magnesium (Mg) and Silicon (Si) peaks have very good peak background ratios. It can be seen that the spectrum contains small peaks of iron (Fe), Sulphur (S) and Aluminum (Al). The higher peaks of the iron are from the metal (nickel) grid used to support carbon film.

As a general observation, the major constituents of a sample can usually be identified with a high degree of confidence. Comparing the characteristics of all types of asbestos with their chemical composition from chapter 2 and appendix 5, the results shown above counterparts to amphibole type's tremolite ($Ca_2Mg_5Si_8O_{22}(OH)_2$) and actinolite ($Ca_2(Mg,Fe)_5Si_8O_{22}(OH)_2$).

3.5.1.2.3 Fiber length and width

The fibrogenicity and carcinogenicity of asbestos fibers are dependent on several fiber parameters including fiber dimensions.

SEM images of the sample observed were quantified with the programs ImageJ and Gimp2.0, and with the help of these software, length and width of single fibers and also of bundles were measured along with the grain size present in the sample.

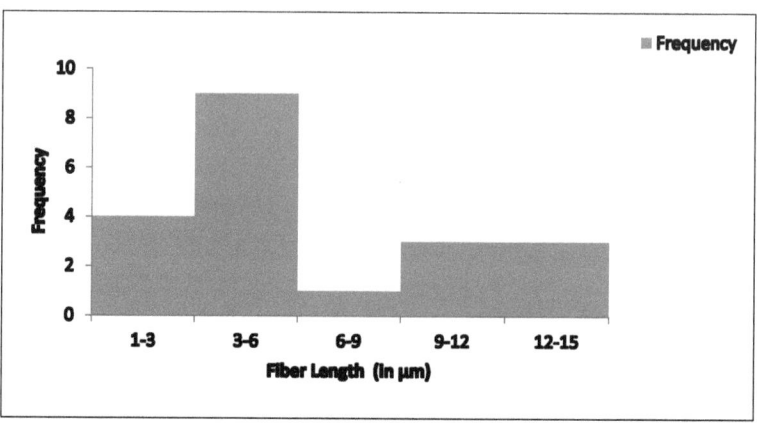

Figure 13: Histogram showing Fiber length in μm (These single fiber length quantification is carried from samples MY_1-2-3_SE and MY_1-1-4_SE (Appendix 3)).

More than 50 per cent of detected samples were longer than or equal to mean value of 6.6910 μm as shown in the frequency distribution table (Appendix 1).

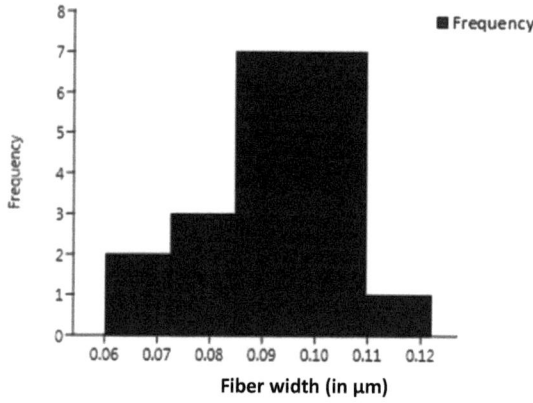

Figure 14: Histogram showing Fiber width in μm (These single fiber width quantification is carried from samples MY_1-2-3_SE and MY_1-1-4_SE (Appendix 3)).

Almost all the asbestos fibers detected in these samples were thinner than or equal to mean value of 0.0922 μm as clearly shown in the histogram above. As shown in the frequency table (Appendix 2), it is concluded that the width of the fibers in the sample ranges from 0.85 to 0.11 μm.

Length and Width of the fibers are both important parameters in determining the carcinogenic potential of asbestos and other specific fibrous materials.

According to Bureau of Indian Standard-BIS: 11450 (2006) and World Health Organization (Boulanger et al., 2014) current regulations focalize as per following fiber definition:

Fiber length >5 µm and Fiber width <3 µm, Aspect ratio > 3:1

Only fibers thinner than 3 µm, longer than 5 µm and a length/width ratio above 3 are taken into account for regulatory purposes. Shorter asbestos fibers with length less than 5 µm, width more than 3 µm are not taken into consideration. According to asbestos cement sample analysis, asbestos fibers were detected with the width less than 0.13 µm and length more than or equal to 5 µm. This comprise that the shorter fibers less than 5 µm and thinner fibers are easily respirable and more vulnerable to asbestos-related diseases.

In **India**, according to Model rule 123-A under section 112 of the factories Act [Van Zandwijk et al., 2013], exposure limit to various types of fibers are defined as follows:

Model Rule 123-A under section 112 of the Factories Act	
Amosite	0.5 f/ml
Chrysotile	2.0 f/ml
Crocidolite	0.2 f/ml
Other forms	2.0 f/ml

Short fibers are more easily phagocytized by alveolar macrophages than long fibers, so their retention half-life is shorter, and long fibers are generally found to be more persistent in the lung than shorter fibers. Consequently, regarding the size dependent biological effects of asbestos, it seems that the content of both longer and shorter fibers present in the air should be taken into consideration (Boulanger et al., 2014). Although it is common to see the dimensions of asbestos fibers discussed in terms of the ratio of length to width, or aspect ratio, the use of such a dimensionless parameter result in the loss of information about the size of fibers and therefore is of little use in the discussion of fiber carcinogenicity or exposure (Wylie et al., 1993). Asbestos fibers of all lengths induce pathological responses and that caution should be exerted when attempting to exclude any population of inhaled fibers, based on their length, from being contributors to the potential for the development of asbestos-related diseases [Dodson et al, 2003].

One more factor needs to be considered while determining the carcinogenic potential of asbestos i.e. fiber dimension along with the physical structure of asbestos fiber. Fiber physical characteristics have been identified as critical in manipulating the pathogenesis of fiber-related adverse respiratory effects. Fiber dimension acts as an important role in determining how these fibers penetrate lung tissues, meaning capable of being respirable. When damage of asbestos-containing materials, asbestos tends to split up longitudinally into new fine, straight fibers: these fibrils are much thinner, more respirable, and as a result more dangerous than main fibers.

Some of the fibers having carcinogenic potential are listed below:

1. Thick straight fiber bundle with blunt end

2. Straight thick bundle with sharp and pointed fiber ends

3. Thick fiber bundle with short curvy ends

4. Thin fiber bundle split up with number of long curvy ends

5. Thin straight fiber bundle with blunted end

6. Single straight fiber

7. Single straight fiber with curvy end

8. Single curvy fiber

9. Curved thick fiber bundle attached with small grains to its end

10. Single fiber with small grains attached to its end

Above mentioned different fiber shapes develop when any asbestos-containing material is forcibly broken into smaller pieces. Bundled fibers can be found attached to other materials as well as split up with number of smaller pointed, sharp fibers or curvy fibers. Likewise single fibers can also be found in various shapes. For example, Fiber shape 7 can be the result from fiber shape 8 in its stable state while as it is well known that fibers are flexible and elastic in nature. Asbestos fibers are not friable or airborne when it is attached with other materials as shown above in fiber shape 9 and 10. Hence, 9 and 10 fiber shapes are not inhalable.

Even though fiber dimension along with physical structure is linked to the pathogenic effects of asbestos, it is also recognized that fiber characteristics like durability, harshness, surface chemistry, surface area or activity, etc.) are likely to play an important role in the pathogenic process (Boulanger et al., 2014).

For getting information about the orientation of the fibers in the layers of asbestos cement AMOCADO analysis were performed. AMOCADO modular software tool, conducted in the MATLAB environment to quantify the anisotropy of a pattern's complexity [Gerik and Kruhl, 2009]. The Anisotropy analysis in AMOCADO requires digitized image of the SEM image, which is processed into a binary format (i.e., black and white pattern).

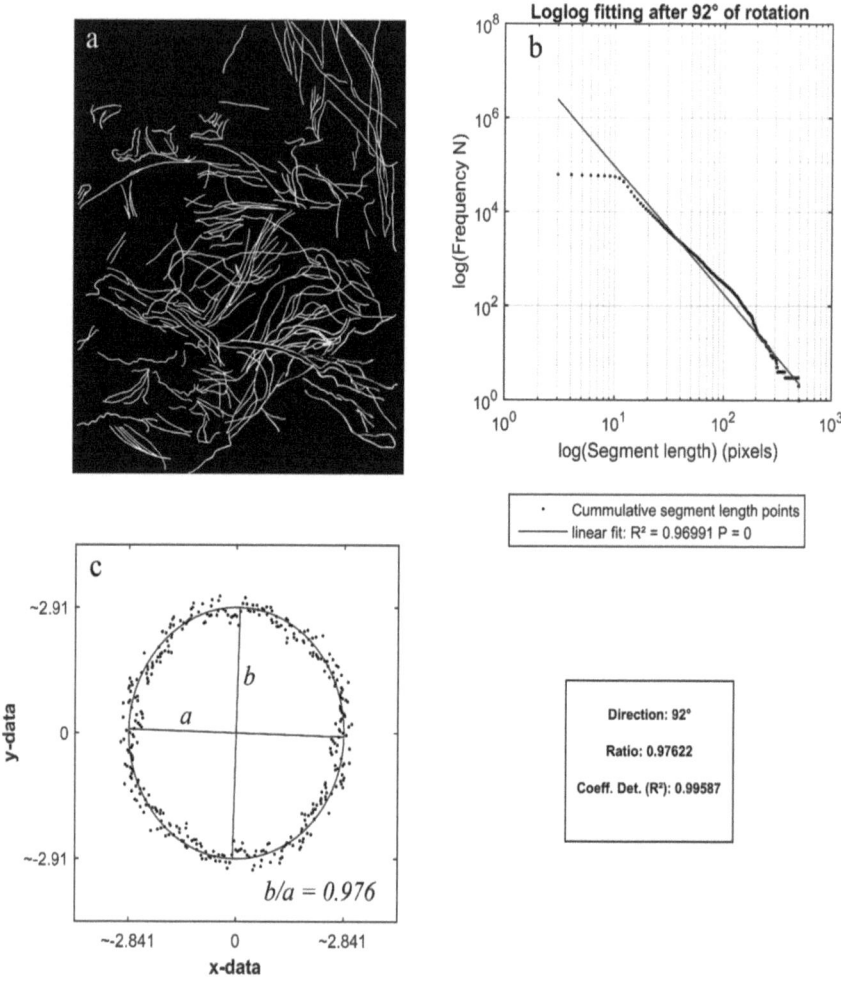

Figure 15: Analysis of asbestos fiber to determine its fiber complexity. (a) Binary image of SEM image MY1-1-2.tif. (One pixel = 0.00772 μm) (b) Basic diagram of the modified Cantor-dust method (AMOCADO software, Gerik and Kruhl, 2009), exemplified for the 92° direction (figure a). Number of segments with length N is plotted cumulatively against segment length s on the log-log diagram. Red line shows corresponding linear regression line with slope m. The slope value m is -2.90. The slope is determined based on the power law relationship between the length of the segments vs number of segments in a double log plots. (c) Representation of m-values for all 180 directions (with 1° angular distance) through the branching structure. The m-values are plotted from the center to the outside. Spatial arrangement of data points is estimated by fitting the ellipse with axis 'a' and 'b', where 'a' indicates long axis and 'b' indicates short axis.

Modified Cantor-dust method (MCDM) using AMOCADO software (Gerik and Kruhl, 2009), and intercept method (Launeau et al., 2010) are applied to quantify the intensity of pattern anisotropy of the asbestos fiber. The linear arrangement of segment length vs. cumulative frequency data points in the double logarithmic plot indicates self-similarity, i.e. fractality, of the pattern over the entire range of measurement, characterized by the slope value m. The best-fit ellipse highlights the geometry of the data point arrangement and serves as a measure for the anisotropy of the structure, given by the ratio of the ellipse axes b and a. Anisotropy intensity is measured based on the ratio of short (b) and long axis (a) of the fit ellipse.

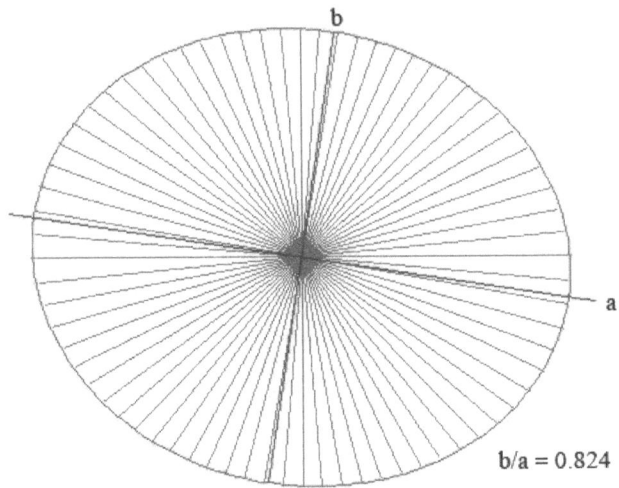

Figure 16: Intercept ellipse with axial ratio

In intercept analysis, the direction dependence of the fibers arrangement shows very weak anisotropy, shown in figure 16 with an axial ratio of 0.824 of the best-fit ellipse through the data points. This very weak anisotropy observes in the direction 98.59° (along a-axis). Interestingly, AMOCADO analysis also gives the similar result, i.e. very weak anisotropy (0.976) shown in figure 15. Such extremely low anisotropic values indicate absence of any direction related processes.

4 HEALTH IMPACTS DUE TO ASBESTOS EXPOSURE IN INDIA

4.1 Toxicity of various types of Asbestos

Asbestos (actinolite, anthophyllite, amosite, crocidolite, tremolite and chrysotile) has been classified by the International Agency for Research on Cancer as being carcinogenic to humans [International Agency for Research on Cancer, 2012]. Exposure to solely chrysotile, amosite and anthophyllite and to mixtures containing crocidolite results in an increased risk of lung cancer [Nat.Toxicology Program, 2011]. Scientists from the United States National Institute for Occupational Safety and Health (NIOSH) concluded that chrysotile asbestos should be treated with virtually the same level of concern as the amphibole forms of asbestos [Dodson and Hammar, 2005; NIOSH, 2011]. Chrysotile asbestos has also produced tumors in animals and is also a recognized cause of asbestosis and malignant mesothelioma in humans [Kanarek, 2011]. Likewise mesothelioma has been observed in people who were occupationally exposed to amosite, tremolite, crocidolite and chrysotile, as well as family members of the occupationally exposed, and also to the general people living in the neighborhood close to asbestos factories and mines [Marbbn, 2009].

4.2 Asbestos Related Disease (ARD)

Fine fibers are more likely to be breathe in than coarse fibers with the reason that they remain suspended in the air for longer. The thin fibers are generally generated when asbestos-containing material is roughly handled, can penetrate deep into the lung where they can lead to potentially life-threatening lung and abdominal conditions, including asbestosis, lung cancer and mesothelioma [The Occupational Safety and Health Administration, 2012]. Exposure to a quantity of airborne asbestos fibers is detrimental to human health. The toxicity of asbestos has been known since the early twentieth century and linked almost exclusively to pleural mesothelioma [McCormack, 2012]. Asbestos are generally 50 to 200 times thinner than a human hair [Wachowski and Domka, 2000], and if breathed into the lungs the possible health effects includes mainly three deadly diseases, characterized by extended latency periods (length of the time between exposure and the onset of diseases) [Allen, 2010]. In India, the latency period is estimated to be 20 to 37 years [Ramanathan and Subramanian, 2001]. Moreover, due to this long latency period for asbestos diseases, diagnosis is difficult, and periodic checkups are almost non-existent. Hence, most of the asbestos-related diseases are not diagnosed [Pandita, 2006]. There are no comprehensive data accessible to analyze the over-all number of deaths in India due to asbestos, since the records do not classify different causes of respiration problems [Subramanian and Madhavan, 2005] and there is no register for ARD.

4.2.1 Asbestosis

Asbestosis is diffuse pulmonary fibrosis caused by the inhalation of asbestos fibers, generally after high level, long term exposure over 15-20 years [Coleman and Tsongalis, 2009]. Fibrosis begins sub-pleural in the lower lobes and may develop extensive after exposure to asbestos has terminated. There is no effective treatment for asbestosis [Greenberg et al., 2003]. Asbestosis tends to be linked to heavy occupational exposure of asbestos fibers. The risk of asbestosis is minimal for persons who are not directly associated with asbestos workings, meaning the disease is rarely caused by neighborhood or family exposure. Those who renovate or demolish buildings that contain asbestos are at significant risk, depending on the nature of the expose and precaution taken [Allen, 2005]. The higher the exposure, the greater the chances of developing asbestosis and the shorter the time it takes [Allen, 2010].

4.2.2 Malignant Mesothelioma

Malignant Mesothelioma is a rare form of cancer which often occurs in the thin membrane lining of the lung, abdomen, chest and (rarely) heart [Asbestos Diseases Research Institute, 2013] and is considered to be one of the most serious of all asbestos-related diseases.

Malignant mesothelioma has generally three most recognized varieties:

- Malignant pericardial mesothelioma
- Malignant pleural mesothelioma, and
- Malignant peritoneal mesothelioma.

Pleural mesothelioma occurs in the pleura, the lung layer. Peritoneal mesothelioma occurs in the peritoneum, the abdominal cavity wall. Pericardial mesothelioma occurs in the lining of the heart, known as the pericardium [Allen, 2010]. The risk of Malignant mesothelioma is high to the people who are close to asbestos mines, mills, factories and shipyards as well as people who install asbestos products, close to asbestos mining areas, product factories where usage of asbestos has large quantities of friable asbestos fibers. Mesothelioma may occur in the absence of asbestosis and can be associated with relatively low exposures to asbestos [Allen, 2010]. It accounts for the majority of victims who contract an asbestos-related disease through environmental exposure and is a notoriously aggressive cancer with no known cure [Allen, 2004]

4.2.3 Asbestos Related Lung Cancer (Bronchial carcinoma)

Bronchial carcinoma can occur from both occupational and non-occupational exposure. It is the predominant malignancy contracted by the release of asbestos fibers. Lung cancer causes the largest number of deaths related to asbestos exposure. The incidence of lung cancer in people who are directly involved in the mining, milling, manufacturing and use of asbestos and its products is much higher than in the general population [Allen, 2007]. There is a powerful synergistic interaction between asbestos exposure and cigarette smoking in the induction of the condition [Allen, 2010]. Those who smoke cigarette have significantly greater risk of suffering lung cancer than people who have only been exposed to asbestos.

Although mesothelioma and other asbestos related cancer like ovarian and laryngeal are recognized around the world, in India neither one of these diseases is commonly reported as cancer a notifiable disease [Kamat, 2008]. There is underreporting in India about asbestos related conditions. Mortality rates of asbestos-related disease is already declining in Western developed countries as a result of bans on asbestos use and can be expected to decline more in the coming years. By contrast, however, there is every expectation that rates of these diseases will rise in countries where asbestos use continues or is increasing, and most of these are low- and middle-income countries [Allen, 2007]. Based on knowledge of past and current exposures to asbestos in industry of India, it can be predicted that the future occurrence of clinical asbestos-related diseases-pleural changes, bronchogenic carcinoma, pulmonary fibrosis, and diffuse malignant mesothelioma will increase. These cases of asbestos-related disease are predictable to occur in asbestos exposed workers from mining, milling, and manufacturing as well as with secondary exposures to asbestos-containing materials, including construction and maintenance workers, consumer of asbestos-containing products, and the tenants of asbestos-containing buildings [Dave and Beckett, 2005].

Tracing the records of asbestos disease has deliberately made difficulties in most of the countries by insufficient disease investigation and by the absence of authorized government. For instance, in **India**, not any official recognition of mesotheliomas was recorded, but one hospital in Mumbai alone has recognized more than 30 cases of mesothelioma treated at that hospital in a single year. As is classically appreciated in public health, the lack of data does not mean the lack of disease [Park et al., 2011].

4.3 Occupational exposure in India

In India, significant occupational exposure to asbestos occurs mainly in asbestos cement factories, asbestos textile industry, asbestos mining and milling, inadequacy compliance to mines and safety act low content of asbestos fiber in parent rock, insulating materials, railway, repair/maintenance, and obsolete technology [DTE, 2000]. There are 18 asbestos cement factories located in different parts of the country. The National Institute of Occupational Health (NIOH) recently carried on environmental epidemiological studies in four asbestos cement factories located in Ahmedabad, Hyderabad, Coimbatore and Mumbai. The prevalence of asbestosis in these cement factories varied from 3 to 5 per cent. Again on milling and mining sector in Andhra Pradesh and Rajasthan, the levels of asbestos fibers were found to be higher than the permissible levels of 2 fibers/ml. The average fiber levels in milling units varied from 45 fibers/ml to 244 fibers/ml of air. The overall prevalence of asbestosis in mining and milling was 3 and 21 per cent respectively. [NIOH, 2006; Ban Asbestos India, 2007]

Conventionally, occupational exposures to those who handle asbestos directly have been reported to be the most hazardous, and asbestos workers develop the maximum quantity of asbestos-related disease [Newhouse and Thompson, 1965]. The possibilities for exposure to asbestos range extreme far away from the place of work. Additional groups at risk include those who experience "bystander" exposure to asbestos [Frank and Joshi, 2014]. These include personnel who are not themselves asbestos workers, but who work in the vicinity to poorly organized asbestos work and are exposed to friable asbestos fibers. Such exposures have been particularly prevalent in the construction occupation as well as in the shipbuilding business. It was suggested that workers should be allowed to take a bath and change their outfit before going home with asbestos on their outfit [Harries, 1968].

In India, workers remain unacquainted and unaware of the vulnerabilities they are experiencing; there is no execution of health and safety regulations in the asbestos sector or in construction industry. Indian researchers have reported numerous cases of high exposure levels to asbestos fibers in the workplace, which indicates a potential epidemic-like situation of asbestos-related diseases in the coming years. [Allen, 2005]

5 INDIAN REGULATION ON ASSESSMENT AND MANAGEMENT OF ASBESTOS EXPOSURE

Removal of asbestos-containing building materials remains a big challenge and an unacknowledged problem in India. Management of asbestos-containing waste is seriously an issue. Once asbestos-containing material is placed in buildings and infrastructure in communities, it stays there for decades, causing harm to health of countless men, women and children, particularly as it deteriorates, is broken, is cut, gets re-used for other purposes, gets dumped. There is an immense problem with repairing, renovation and maintenance of asbestos products as these operations causes dust release from asbestos-cement products while cutting, drilling, breaking, sanding, grinding, filing, and dismantling. Asbestos roofs and pipes are mostly being removed from various places and agencies in an unscientific way and disposed openly to municipal landfills, creating complications even for the general public in the vicinity. It is truly a deadly ticking time bomb that continues for years and years to create more victims. The cost of removing damaged and broken asbestos products, or remediation of damaged asbestos products, is enormous and represents a huge economic burden for future generations.

5.1 Waste Handling

Employers should handle asbestos-containing waste in a cautious manner so that it does not possess any health risk to the workers concerned or the residents in the vicinity. Disturbing asbestos-containing materials during construction or after construction for renovation is a serious hazard. To control exposure to asbestos in the workplace, building owners and asbestos-containing materials technician's need to wear Occupational Safety and Health Administration (OSHA) compliant respirators and Personal Protective Equipment (PPE) [OSHA, 2002; OSHA Technical Manual].

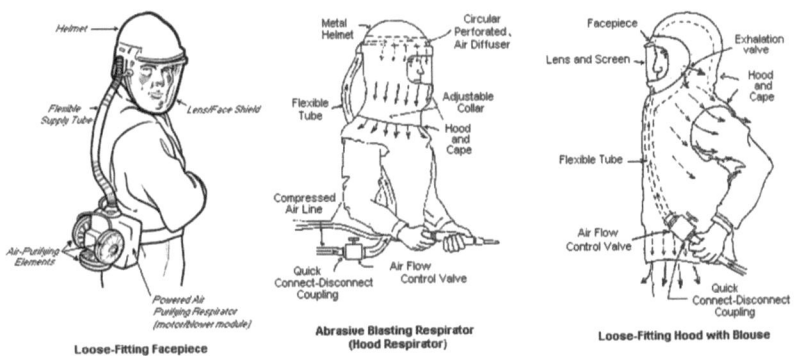

Figure 17: OSHA compliant respirators and personal protective equipment [OSHA Technical Manual]

Indian Standard 11768-1986 (Reaffirmed 2005) has provided recommendations for disposal of asbestos waste material without significant generation of airborne asbestos fibers. The provisions of this standard is applied to any place where any form of asbestos waste is generated, stored, transported and finally disposed of.

General requirements while dealing with asbestos-containing materials as Bureau of Indian Standard-BIS 11768-1986 (Reaffirmed 2005):

- Every single employer who undertakes work which is accountable to generate asbestos-containing waste, should take adequate steps to prevent the generation of friable asbestos dust during handling, storing, transportation and final disposal of asbestos-containing materials.
- Containers collecting waste itself shouldn't be contaminated with any asbestos dust from outside as well as it should be air-tight.
- Before disposal, asbestos waste containers should be securely closed and stored in a separate storage area where they are unlikely to be damaged by external forces. Most importantly, it shouldn't be mixed with containers of other materials including waste materials.

5.1.1 Waste Avoidance

The best way to tackle the solid waste is by following waste avoidance. The most desirable method of controlling waste is to minimize the amount of waste that is generated. This can be generally done by adopting, choosing a process or operation that reduces generation of waste and or recycling of the waste.

However, asbestos can be recycled by changing it completely into nontoxic silicate glass. A procedure of thermal decomposition at 1000–1250 °C produces a combination of non-hazardous silicate phases, then at temperatures above 1250 °C it produces silicate glass [Gualtieri and Tartaglia, 2000]. Microwave thermal analysis can be used in manufacturing process to change asbestos and asbestos-containing material waste into porcelain stoneware tiles, porous single-fired wall tiles, and ceramic bricks [Leonelli et al., 2006].

5.1.2 Waste Collection

According to Bureau of Indian Standard 11768-1986 (Reaffirmed 2005), collection of asbestos waste should be done by a suitable and effective method without releasing any dust, loose fiber, off-cuts broken pieces, bagged asbestos and wet waste to the atmosphere.

Dust: Bagging of outlets from dust collection hoppers should be designed to make bag-changing easier and to minimize dust leakage. Bags of translucent material such as polyethylene should be used wherever practicable so that the dust level can be seen and overfilling can be avoided. Water-soluble paper sacks should not be used where any risk exists of deterioration by wetting before final disposal. Suitable protective clothing and respirators should be worn when bags on a dust collector are changed.

Loose fiber and floor sweepings: Loose fiber handled by fixed extraction systems should, wherever practicable, be returned to the production process. Floor sweeping should be cleaned with a suitable vacuum cleaner and should be placed in airtight sealed bags.

Off-cuts, Broken Pieces and Rejects of High Density Materials: Hard waste, such as bonded asbestos, asbestos cement, jointing's and rubber residues, etc. should be stored in such a way as to confirm that it will not be crushed while awaiting disposal.

Sacks or Bags Which Have Contained Asbestos: Sacks or bags which have contained loose asbestos fibers should be disposed of by grinding, melting or bagging and these used sacks or bags should be collected under strict dust control conditions in impermeable containers, such as unused plastic bags, in addition such containers should be closed and sealed.

Wet waste: asbestos sludge or slurry: Asbestos waste in the form of sludge should preferably be recycled or loaded into particularly designed carriers or other waste containers in such a way as to confirm that no spillage which may subsequently dry out occurs.

5.1.3 Identification and Isolation of waste

Once all asbestos-containing waste material has been identified and isolated, it must be cautiously inspected for damage as Bureau of Indian Standard 11768-1986 (Reaffirmed 2005). If any damage occurs in the building, building owners and employers must resolve on the harmless method, based on the use of the asbestos, quantity of damage, potential for asbestos fibers to be airborne (forming an exposure hazard), etc.

Asbestos waste awaiting disposal should firstly be identified by markings on the sacks or bag and then stored in such a way that it is not exposed to damage likely to cause spillage. Asbestos waste should not be mixed with other waste for which there are no special disposal requirements. Where practicable, a special area should be set aside for its storage.

5.1.4 Transportation of Asbestos waste

Asbestos waste, whether loose or in sealed containers, should be transported to the disposal point in such a way that no asbestos dust is emitted into the air during transport. In case of accidental spillage, it should be wetted if practicable and covered immediately. The material should then be removed and during this process appropriate safety precautions, which may include the use of protective clothing and respiratory equipment, should be taken and should be issued to drivers of vehicles carrying asbestos waste stated by Bureau of Indian Standard 11768-1986 (Reaffirmed 2005).

5.1.5 Disposal of asbestos waste

Before a site is used for the disposal of asbestos waste, care should be taken to establish that the site is both suitable and acceptable for the purpose. The disposal site chosen should have vehicular access to the working face, or to a hole or trench dug to receive the asbestos waste. According to Bureau of Indian Standard 11768-1986 (Reaffirmed 2005), the waste should wherever practicable be deposited at the foot of the working face of the landfill site or at the bottom of an excavation dug for it. Where the waste has to be deposited from above the working face, or into an excavation, care should be taken to prevent spillage from bags. When deposited, all waste other than high-density waste should be covered to an acceptable depth (for example 8-10 inch) as soon as possible. No asbestos waste should be left uncovered at the end of a working day. Final covering of asbestos waste should be to a minimum depth of 6 feet. 6 inch as Bureau of Indian Standard. Deposited damp waste must be shielded in the similar method as dry waste to prevent the escape of asbestos dust on drying out. Wet pits should not normally be used for the disposal of any asbestos waste other than high-density material. Where high-density waste is deposited on a dry site, care should be taken to ensure that it is not ground to dust by the passage of vehicles over it. Warning sign should be clearly displayed at the landfill site.

5.1.6 Personal protection and hygiene

Workers engaged in the handling of asbestos waste possibly are at risk of exposure to friable asbestos fibers should be handover appropriate protective outfit and breathing equipment. Where vehicles and reusable containers and covers have been in contact when handling asbestos waste, they should also be cleaned after use by means of a vacuum cleaner or by an alternative dustless method stated by Bureau of Indian Standard 11768-1986 (Reaffirmed 2005).

5.1.7 Supervision

Periodic supervision should be undertaken to ensure that the necessary safety precautions are being followed. If a waste disposal contractor is employed, the relevant sections of the code should be incorporated in the contract. The agreement should formally state that the service provider is accountable for confirming the safety measures are observed at the disposal site. Periodic checks should also be made by the undertaking to ensure that the contractor is observing the code. Where an undertaking disposes of its own asbestos waste, written instructions should be issued to the workers concerned. Besides, what has already been stated by Bureau of Indian Standard 11768-1986 (Reaffirmed 2005), appropriate measures should be taken to prevent pollution of the soil, subsoil, air and water.

6 SAFER ALTERNATIVES OF ASBESTOS

Asbestos substitutes are manmade mineral fibers that have properties similar to those of asbestos and are used to replace asbestos in manufactured products [Virta, 2002]. Manmade mineral fibers are normally produced as fibers of diameter higher than asbestos, and too large in diameter to be respirable [Foa and Basillico, 1999]. To replace asbestos, a wide variety of substitutes have been developed that can be fibrous or non-fibrous, organic or inorganic, and natural or synthetic; each has its own unique physical and chemical properties.

In 2003, approximately 85 per cent of world consumption accounted for asbestos cement products [Virta, 2005]. Asbestos-containing sheets and pipes are produced to be used as economical construction materials in about 100 countries [Tossavainen 2004]. Conversely, these asbestos-containing materials can be substituted with ductile high-density polyethylene pipe, iron pipe, and metal wire reinforced solid pipe.

Differing from asbestos substitutes, alternative products substitutes an entire asbestos-containing material rather than just the asbestos used within the product. For example, cast-iron and PolyVinyl-Chloride (PVC) pipes are alternative products that are used in place of asbestos-cement pipe [Kogel, 2006; Castleman, 2009]. PVC also poses health risks, but not as severe ones as asbestos. Alternatively, fibrous cement can be completely substituted by using metal or plastic to form the section. It is likely to eliminate the usage of asbestos by redesigning the product, or by using another safer non-hazardous material. Multiple manufacturers create safer alternative these days. The substitute products consist of fiber-cements made with polymeric and plant fibers, as well as wholly altered product configurations that function the same functions as asbestos-cement corrugated sheets, pipe, and water tanks [WHO, 2012].

Because of the widespread exposure and an increasing movement away from consuming asbestos-containing materials, finding an appropriate substitute to asbestos products turned out to be principal for many manufacturers globally. For each product there are man-made safer alternatives. The most common are polyurethane foam, amorphous silica fabric, thermoset plastic flour, flour fillers and cellulose fiber. Alternative products are practical for many applications, because they usually are existing products and do not require the manufacturer to develop a new asbestos-free product. The major concerns of customers and manufacturers when selecting an alternative product are cost and performance [Hodgson, 1985]. As a result, when renovating a home or an office building, there are many varieties to be measured among which many of the alternative products are "sustainable", providing a safer alternative for the environment and most prominently for the individuals living in the building. Almost all the asbestos products can be substituted sustainably, which is clearly proven by the fact that many countries banned asbestos without any difficulties replacing asbestos-containing materials.

6.1 List of asbestos alternatives

Safer alternatives for asbestos-containing products of all varieties is increasingly available in the market. Substitutes for asbestos products are not limited to products that simply replace asbestos with other material (e.g., polyvinyl acetate and cellulose in fiber-cement corrugated sheet). Furthermore, there are also a number of exclusively altered products that can substitute the asbestos products. A number of substitutes for asbestos products are included in the following table:

Asbestos Product	Asbestos substitute or alternative product
Flooring	Carpeting, ceramic tile, clay, fiberglass, polyethylene pulp, silica, talc, vinyl compositions, wood
Asbestos-Cement Corrugated Roofing	Fiber-cement roofing using: synthetic fibers (polyvinyl alcohol, polypropylene) and vegetable/cellulose fibers (softwood Kraft pulp, bamboo, sisal, coir, rattan shavings and tobacco stalks, etc.); with optional silica fume, fly ash, or rice husk ash Micro-concrete (Parry) tiles Galvanized metal sheets Clay tiles Vegetable fibers in asphalt Slate Coated metal tiles (Harveytile) Aluminum roof tiles (Dekra Tile) Recycled polypropylene and high-density polyethylene and crushed stone (Worldroof) Plastic coated aluminum Plastic coated galvanized steel.
Insulation	Calcium silicate board, cement board, ceramic fiber, fiberglass, mica, mineral wool, vermiculite
Textile	Aramid fiber, carbon fiber, ceramic fiber, fiberglass, mineral wool, polybenzimidazole fiber
Asbestos-Cement Flat Sheet (ceilings, facades, partitions)	Fiber-cement using vegetable/cellulose fibers (see above), wastepaper, optionally synthetic fibers

	Gypsum ceiling boards (BHP Gypsum) Polystyrene ceilings, cornices, and partitions Façade applications in polystyrene structural walls (coated with plaster) Aluminum cladding (Alucabond) Brick Galvanized frame with plaster-board or calcium silicate board facing Softwood frame with plasterboard or calcium silicate board facing.
Plastics	Aramid fiber, carbon fiber, fiberglass, fumed silica powder, mica, polytetrafluoroethylene, potassium titanate, wollastonite
Friction	Aramid fibers, cellulose, ceramic fiber, fiberglass, metal (brass, bronze, copper, iron) fibers, palygorskite (attapulgite), polyacrylonitrile fiber, potassium titanate, semimetallic brakes, sepiolite, steel fibers, vermiculite, wollastonite
Asbestos-Cement Pipe	High Pressure: Cast iron and ductile iron pipe High-density polyethylene pipe Polyvinyl chloride pipe Steel-reinforced concrete pipe (large sizes) Glass-reinforced polyester pipe Low Pressure: Cellulose-cement pipe Cellulose/PVA fiber-cement pipe Clay pipe Glass-reinforced polyester pipe Steel-reinforced concrete pipe (large diameter drainage)
Asbestos-Cement Water Storage Tanks	Cellulose-cement Polyethylene Fiberglass Steel Galvanized iron PVA-cellulose fiber-cement
Asbestos-Cement Rainwater Gutters; Open Drains (Mining Industry)	Galvanized iron Aluminum Hand-molded cellulose-cement PVC
Coatings and compounds	Aramid fiber, carbon fiber, cellulose fiber, clay,

	fiberglass, polyethylene films, limestone, rubber membrane roofing, mica, polyethylene fiber, polypropylene fiber, talc, wollastonite

Table 4: List of Asbestos Alternatives [Castleman, 2009; WorldBank Group, 2009; Virta 2005; WHO, 2012]

Numerous substitutes are available for roofing, interior decoration on walls and ceilings, together with fiber-cement tiles and corrugated sheet products that are manufactured using polyvinyl alcohol fibers and cellulose fibers. Almost all of the polymeric and cellulose fibers used as an alternative of asbestos in fiber-cement sheets products are beyond 10 μm in diameter and consequently it's not inhalable [WHO 2006]. For roofing or slating, light weight tiles can be used in the most rural areas using locally available plant fibers, such as jute, hemp, sisal, palm nut, coconut coir, and wood pulp. Clay tiles and galvanized iron for slating or roofing are amongst other substitute products [World Bank Group, 2009].

Non-asbestos fiber-cement sheets are lighter, less fragile, and have better-quality over asbestos cement corrugated sheets. Compared with asbestos-cement pipes, iron pipes can be transported with less effort and damage, take greater firmness loading, and last longer [World Bank Group, 2009]. When alternative materials for asbestos are being chosen, major attention should be given to related health hazards; and concern should also be paid to the technological and economical conditions which conclude the necessity of alternative products.

7 RESULTS AND DISCUSSION

This chapter presents the results and discussions of the quantitative and qualitative data analysis and their relation to the core objectives and literature review.

Many researchers pointed 90 per cent of asbestos fiber comprised in asbestos cement is chrysotile asbestos [WHO, 2014b; Guidotti, 2011; Dyczek, 2004]. In the sample studied, SEM with EDXA greatly increased the potentiality to confidently classify asbestos fibers and distinguish among the different types of asbestos fibers (Appendix 5). Specific area measured in the SEM image (figure 9) had elemental peaks consistent with tremolite asbestos and another had elemental peak consistent with actinolite asbestos measured from SEM image (figure 11). As a result, the hypothesis that only chrysotile asbestos is present in the asbestos cement was not supported by the data analysis. Other asbestos fiber types are also present in the asbestos rather than only chrysotile asbestos.

Length and width of the asbestos fibers were also measured to know if the sample contained any friable and inhalable fibers. From the SEM image (Appendix 3), length and width of the asbestos fibers were measured using GIMP 2.0 program and were detected. Width of the asbestos fiber were lesser than 3 µm i.e. 0.13 µm and length more than 5 µm, which concluded that asbestos fibers present in asbestos cement is friable and is easily inhalable and takes a shorter time period to phagocytized by alveolar macrophages than longer fibers (>5 µm). It does not mean that the long fibers are not hazardous to human health. Longer fibers are more carcinogenic and persistent as it is piled up in the lung for a longer time period solubilizing with chemical elements inside and then slowly breaking them into smaller fibers. Hence, exposure to fibrils with various length and width, both induce carcinogens.

Various shapes of asbestos fibers were develop when mechanically damaging asbestos cement. As the shape of fibers plays an important role in asbestos-related disease, it was found that friable curvy, sharp, and needle like fibers can easily penetrate lung tissues and induce cancer. It was also determined that asbestos fibers do not possess any health risk, until and unless it is sealed and attached to other materials, meaning non-respirable.

With AMOCADO analysis, anisotropy of the pattern's complexity of the fibers in the layers of asbestos cement was analyzed. From the analysis, it showed the direction dependence of the fiber arrangement has very weak anisotropy. Such extremely low anisotropic values (i.e. axial ratio=0.976) indicates the absence of any specific direction related process. The axial ratio of the fit ellipse can be taken as a measure of the anisotropy intensity and as a result, it represents an important parameter of fiber quantification. Above all, the asymmetrical point distributions and more precisely, the deviations from elliptical arrangements point to a variability of the pattern anisotropy.

8 CONCLUSION

Asbestos is used in India purely for financial reasons. Mining of asbestos has been banned for a long time in India, but the manufacturing of products is still going on. The official reason given by both industry and government is that it is a material for the poor as it is cheap and has many good properties.

As asbestos is primarily used in India for roofs and pipes with a small quantity used for brake linings and other products. Safer substitutes for roofs and pipes are already available with some companies even manufacturing both asbestos and asbestos free products also for the health conscious. Same is true for brake manufacturers. Many of the alternative materials are not more expensive than asbestos-containing materials.

There are several standards specified under the Factories Act and Mines Act including limits of exposure (Appendix 4). But Standards in India are only on paper. All rules and regulations exist but are hardly followed. Large number of violations have been noticed and have brought it to notice of government agencies. Indian railways are removing asbestos roof sheets from several platforms but the waste is dumped on the platform. The sheets are not removed scientifically causing more harm even to the local passengers.

None are aware of asbestos hazards, but the industry claims to use all measures. Mohit Gupta from Asian Network for the Rights of Occupational and Environmental Victims (ANROEV) and Occupational and Environmental Health Network of India (OEHNI) recently visited a thermal power plant where workers narrated not receiving any protective equipment or any measures being in place [Gupta, 2015]. Same is the story in majority of workplaces. Some companies might be using some measures and give PPE (Personal Protective Equipment) to some workers. Medical checkup might also be done for workers, but this is not comprehensive and the results are not disclosed to the workers and not all workers are covered especially under contract workers.

There isn't any awareness program in India in place. There is supposedly a National Program for Elimination of asbestos-related diseases by Ministry of Labor but what is the status is not clear. This program in itself is in contrast to official government positions.

Factors influencing an ultimate ban on asbestos:

- Public awareness about the hazards of asbestos is very important. There have been several studies done by government agencies which were hugely influenced by the industry (an NIOH study was partially funded by industry) where the results show asbestos to be safe. These studies are used to back the arguments from the industry that asbestos can be safely used. There is a need for independent assessment and study of the risks of asbestos to workers and the community.
- Diagnosis of workers suffering from ARDs is another important aspect. Most of the workers are misdiagnosed and there is a need to augment and strengthen the diagnostic infrastructure. All workers in exposed working environments should be tracked and medical checkup done every year to ascertain the risks and deteriorating health. This has also been ordered by the Supreme Court but hardly implemented in true spirit.
- Substitutes should be encouraged by the government by reducing taxes and duty, but this cannot happen till the time there is an understanding among the decision making officials that asbestos is dangerous.
- Reporting and data gathering system of the government can help if correct data is supplied to relevant agencies so that policy formation can be influenced.

More effort is needed in order to completely ban asbestos in India. For doing so, a progressive industry, academia, UN organizations and civil society have to cooperate more to finally convince the Indian government to introduce and enforce a full ban of all types of asbestos production and use in India, as well as establishing safe waste management.

9 REFERENCES

Allen, L.K. (2004). Asbestos related Diseases. International Ban Asbestos Secretariat. Retrieved from http://www.ibasecretariat.org/lka_ards.php (Accessed on July, 2015).

Allen, L.K. (2005). Asbestos and mesothelioma: Worldwide trends. Lung Cancer; 49: S3–8.

Allen, L.K. (2007). Killing the future—asbestos use in Asia. International Ban Asbestos Secretariat, London, UK.

Allen, L.K. (2010). Asbestos-Related Diseases. International Ban Asbestos Secretariat. Retrieved from http://ibasecretariat.org/prof_ards.php (Accessed on July, 2015).

Allred, M.; Campolucci, S.; Falk, H.; Ganguly, N.K.; Saiyed, H.N.; and Shah, B. (2003). Bilateral environmental and occupational health program with India. Int J Hyg Environ Health; 206: 323–32.

Ansari, F. A.; Bihari, V.; Rastogi, S. K.; Ashquin, M.; and Ahmad, I. (2007). Environmental health survey in asbestos cement sheets manufacturing industry. Indian Journal of Occupational and Environmental Medicine, 11(1), 15–20.

Ansari, F.A.; Ahmad, I.; Yunus, M.; and Rahman, Q. (2005). Asbestos: foe or friend? Indmedica Occupational Health – Cyber.

ATSDR-Agency for Toxic Substances and Disease Registry. (2001). Toxicological Profile for Asbestos. Retrieved from http://www.atsdr.cdc.gov/toxprofiles/tp61.pdf (Accessed July 2015).

Ban Asbestos India. (2007). The powder that chokes. Retrieved from http://banasbestosindia.blogspot.de/2007/11/powder-that-chokes.html (Accessed August 2015).

Berman, D.W.; and Crump, K.S. (2003). Final draft: technical support document for a protocol to assess asbestos-related risk. Washington DC: U.S. Environmental Protection Agency. p. 474.

Biswas, D. (1999). Comprehensive Industry Document on Asbestos Products Manufacturing Industry. Central Pollution Control Board, Series: CO, 16.

Boulanger, G.; Andujar, P.; Pairon, J. C.; Billon-Galland, M. A.; Dion, C.; Dumortier, P.; and Jaurand, M. C. (2014). Quantification of short and long asbestos fibers to assess asbestos exposure: a review of fiber size toxicity. Environmental Health : A Global Access Science Source, 13(1), 59.

BIS-Bureau of Indian Standards: 11769 (Part1) – (1987). Guidelines for safe use of products containing asbestos (Asbestos cement products).

BIS-Bureau of Indian Standards: 11769 (Part2) – (1986). Guidelines for safe use of products containing asbestos (Friction materials).

BIS-Bureau of Indian Standards: 11769 (Part3) – (1986). Guidelines for safe use of products containing asbestos (Non-cement asbestos products other than friction materials).

BIS-Bureau of Indian Standards: 11768-1986 (Reaffirmed 2005). Recommendations for disposal of asbestos waste material.

Bureau of Indian Standards-BIS: 11450. (2006). Method for Determination of Airborne Asbestos Fibre Concentration in Work Environment by Light Microscopy (Membrane Filter Method) [CED 53: Cement Matrix Products].

Castleman, B. (2009). Substitute for Asbestos-Cement Construction Products. International Ban Asbestos Secretariat. Retrieved from http://www.ibasecretariat.org/bc_subst_asb_cem_constr_prods.php (Accessed June 2015).

Castleman, B.I.; and Joshi, T.K. (2007). The Global Asbestos struggle today. Eur. J. Oncol., vol. 12, n. 3, pp. 149-154.

Coleman, W.B.; and Tsongalis, G.J. (2009). Molecular Pathology: The Molecular Basis of Human Disease; Elsevier Science.

Dave, S.K.; and Beckett, W.S. (2005). Occupational asbestos exposure and predictable asbestos - related diseases in India. Am Rev Respir Dis; 48: 137–43.

Dyczek, J. (2004). Surface of Asbestos-cement (AC) Roof Sheets and Assessment of the Risk of Asbestos Release. Technical University of Mining and Metallurgy, Poland. World Asbestos Report.

Dodson, R. F.; Atkinson, M. A.; and Levin, J. L. (2003). Asbestos fiber length as related to potential pathogenicity: a critical review. American journal of industrial medicine, 44(3), 291-297.

Dodson, R.F.; and Hammar, S.P. (2005). Asbestos: Risk Assessment, Epidemiology, and Health Effects; CRC Press.

DTE-Down to Earth. (2000). Death inside the factory (asbestos) gate. Scientific magazine.ed. by Agrawal, A. 38-42, Center for Science and Environment, New Delhi.

Environmental Impact Assessment Guidance Manual (2010). Guidance Manual for asbestos-based industries. Ministry of Environment and Forests, Government of India.

Federal Institute for Occupational Safety and Health. (2014). National Asbestos Profile for Germany. Bundesanstalt für Arbeitsschutz und Arbeitsmedizin (BAuA) Retrieved from http://www.baua.de/dok/5707178 (Accessed June 2015).

Feininger, T. (2003). The Manual of Mineral Science. 647 pages. ISBN 0–471–25177–1. The Canadian Mineralogist, vol. 41 no. 2 542-554.

Foa, V.; and Basilico, S. (1999). Chemical and physical characteristics and toxicology of man-made mineral fibers. La Medicina del lavoro, 90(1), 10-52.

Frank, A. L.; and Joshi, T. K. (2014). The global spread of asbestos. Annals of Global Health, 80(4), 257–62.

Gee, D.; and Greenberg, M. (2002). Asbestos: from 'magic' to malevolent mineral, Late lessons from early warnings: the precautionary principle 1896–2000 (Copenhagen: EEA) (22): 52–63. ISBN 92-9167-323-4.

Gerik,A.; and Kruhl,J.H. (2009): Towards automated pattern quantification: Time-efficient assessment of anisotropy of 2D patterns with AMOCADO. Computers & Geosciences 35/6, 1087-1097, do:10.1016/j.cageo.2008.01.015i.

GIA-Gemological Institute of America (1988). Gem Reference Guide. City: Gemological Institute of America. ISBN 0-87311-019-6.

Graham, A. (2008). Asbestos Identification and Quantification in Bulk Samples. EMLab P&K Asbestos Analyst, The Environmental reporter, Vol 6 Issue 12.

Gravatt, C. C.; Philip, D.; and LaFleur, K. F. J. H. (1978). Proceedings of Workshop on Asbestos, Definitions and Measurement Methods. Department of Commerce, National Bureau of Standards.

Gualtieri, A. F.; and Tartaglia, A. (2000): Thermal decomposition of asbestos and recycling in traditional ceramics. Journal of the European Ceramic Society 20 (9): 1409.

Guidotti, T.L. (2011). Global Occupational Health. Oxford University Press. Retrieved from https://books.google.com/books?id=ih8Pqid2q0UC&pgis=1 (Accessed July 2015).

Gupta, M. (2015). Asbestos in India Abandoned and neglected myths and realities. Asian Network for the Rights of Occupational and Environmental Victims (ANROEV) and Occupational and Environmental Health Network of India (OEHNI).

Harries, P.G. (1968). Asbestos hazards in navel dockyards. Ann Occ Hyg; 11:135-145.

Health and Safety Executive. (2012). Managing and Working with Asbestos. Retrieved from http://www.hse.gov.uk/pubns/priced/l143.pdf (Accessed July 2015).

Hodgson, A.A. (1985). Alternatives to Asbestos and Asbestos Products. Anjalena Publications. P 230 Berkshire, UK.

Hylton, K. N. (2008). Asbestos and Mass Torts with Fraudulent Victims. Sw. UL Rev., 37, 575.

ILO- International Labor organization. (1984). Safety in the use of Asbestos. ILO code of Practice. Geneva, International Labor Office.

ILO- International Labor organization. (1999). ILO Estimates Over 1 Million Work-Related Fatalities Each Year. Geneva, ILO news.

Indian Bureau of Mines yearbook 2011, Mineral review; Asbestos.

Indian Bureau of Mines yearbook 2013, Mineral Review; Asbestos.

IARC-International Agency for Research on Cancer Asbestos. (2012). (chrysotile, amosite, crocidolite, tremolite, actinolite, and anthophyllite). IARC Monogr Eval Carcinog Risks Hum; 100C:219-309.

International Ban Asbestos Secretariat. (2014). Graphics. Retrieved from http://www.ibasecretariat.org/graphics_page_row4.php?n=6#mit1_start (Accessed July 2015).

Jadhav, A. V.; and Roy, N. (2012). Asbestosis: Past voices from the Mumbai factory floor. Indian Journal of Occupational and Environmental Medicine, 16(3), 131–136.

Joshi, T.K.; and Gupta, R.K. (2004). Asbestos in developing countries: Magnitude of risk and its practical implications. Int J Occup Med Environ Health; 17: 179–85.

Kamat, S.R. (2008). Asbestos Related disease in India. India's Asbestos Time Bomb. International Ban Asbestos Secretariat, UK; 55.

Kanarek, M.S. (2011). Mesothelioma from Chrysotile Asbestos: Update. Annals of Epidemiology 21 (9): 688. doi:10.1016/j.annepidem.2011.05.010.

Kogel, J.E.; Trivedi, N.C.; Barger, J.M.; and Krukowski, S.T. (2006). Industrial minerals and rocks: commodities, markets, and uses. 7th ed. Society for Mining, Metallurgy, and Exploration, Inc. Littleton, Colorado.

Launeau, P.; Archanjo, C.; and Picard. D. (2010).Two- and three-dimensional shape fabric analysis by the intercept method in grey levels, Tectonophysics, Vol. 492, pp. 230-239.

Lemen, R.A. (2004). Chrysotile asbestos as a cause of mesothelioma: Application of the hill caution model. Int J Occup Environ Health; 10: 233–9.

Leonelli, C.; Veronesi, P.; Boccaccini, D.; Rivasi, M.; Barbieri, L.; Andreola, F.; Lancellotti, I.; Rabitti, D. and Pellacani, G. (2006): Microwave thermal inertisation of asbestos containing waste and its recycling in traditional ceramics. Journal of Hazardous Materials 135: 149.

Marbbn, C.A. (2009). Asbestos Risk Assessment. The Journal of Undergraduate Biological Studies: 12–24.

McCormack, V.; Peto, J.; Byrnes, G.; Straif, K.; and Boffetta, P. (2012). Estimating the asbestos-related lung cancer burden from mesothelioma mortality. British Journal of Cancer, 106(3), 575–584.

Meurman, L. O.; Pukkala, E.; and Hakama, M. (1994). Incidence of cancer among anthophyllite asbestos miners in Finland. Occupational and Environmental Medicine, 51(6), 421–5.

Mukherjee, A.K.; Rajmohan, H.R.; Dave, S.K.; Rajan, B.K.; Kakde, Y.; and Rao, S.R. (1996). Pollution and its control in asbestos milling processes in India. Ind Health 34 (1), 35–43.

Mukherjee, S. (2012). Applied Mineralogy: Applications in Industry and Environment (Vol. 7). Springer Science & Business Media.

Nat.Toxicology Program. (2011). Report on Carcinogens (12th Ed.); DIANE Publishing Company.

National Bureau of Standards. (1979). Publications of the National Bureau of Standards Catalog. U.S. Department of Commerce, National Bureau of Standards.

National Bureau of Standards. Office of Air and Water. United States. (1977). Methods and Standards for Environmental Measurement: Proceedings of the 8th Materials Research Symposium. Issue 464. U.S. Department of Commerce, National Bureau of Standards.

NIOH-National Institute of occupational health (2006). Asbestosis. Retrieved from http://www.nioh.org/projects/asbestosis.html (Accessed July 2015).

Newhouse, M.L. and Thompson, H. (1965). Mesothelioma of pleura and peritoneum following exposure to asbestos in the London area. Brit J Ind Med; 22:261e9.

NIOSH-National Institute for Occupational Safety and Health. (2011). Asbestos Fibers and Other Elongate Mineral Particles: State of the Science and Roadmap for Research. Revised Edition. Dept of Health and human Services. Retrieved from http://www.cdc.gov/niosh/docs/2011-159/pdfs/2011-159.pdf (Accessed June 2015).

OSHA-Occupational Safety and Health Administration. (Revised 2002). Asbestos Standard for the Construction Industry. Retrieved from https://www.osha.gov/Publications/OSHA3096/3096.html (Accessed August 2015).

Official Webpage of ELCA Laboratories, India. http://www.elcalabs.com (Accessed August 2015).

Official webpage of National Accreditation Board for Testing and Calibration Laboratories, India. www.nabl-india.org (Accessed August 2015).

Official webpage of SGS global network, India. www.sgsgroup.in (Accessed August 2015).

Official Webpage of SPECTRO Group of Companies, India. www.spectrogroup.com (Accessed August 2015).

OSHA Technical Manual (OTM) | Section VIII: Chapter 2: Respiratory Protection. (n.d.). Retrieved from https://www.osha.gov/dts/osta/otm/otm_viii/otm_viii_2.html (Accessed June 2015).

Pandita, S. (2006). Banning Asbestos in Asia, campaigns and strategies by the Asian Network for the right of Occupational Accident Victims (ANROAV). Int. J. Occup. Environ. Health 12(3), 248-253.

Park, E.K.; Takahashi, K.; and Hoshuyama, T. (2011). Global magnitude of reported and unreported mesothelioma. Env Health Persp; 119: 514-8.

Ramanathan, A.L.; and Subramanian, V. (2001). Present status of asbestos mining and related health problems in India-a survey. Ind Health; 39: 309–15.

Russ, J. C. (2013). Fundamentals of Energy Dispersive X-Ray Analysis: Butterworths Monographs in Materials. Elsevier Science. Retrieved from https://books.google.com/books?id=Dg_-BAAAQBAJ&pgis=1 (Accessed September 2015).

Roy, A.; Khanra, K.; and Bhattacharyya, N. (2013). Asbestos : A potential food contaminant and associated safety risks to consumers, 3(1), 241–243.

Speranskaya, O.; Tsyguleva, O.; and Astanina, L. (2008). Asbestos: Realities, Problems and Recommendations in cooperation with WECF-Women in Europe for Common Future.

Sreedhar, R.; and Alag, N. (2014). Annual Report 2013-2014. Environics India Retrieved from http://www.environicsindia.in/docs/annual_reports/Annual_Report%202013-2014.pdf (Accessed July 2015).

Subramanian, V.; and Madhavan, N. (2005). Asbestos problem in India. Lung cancer, 49, S9-S12.

The Occupational Safety and Health Administration. (2012). Health Hazards in Construction Workbook; Construction Safety Council.

Tossavainen, A. (2004). Global use of asbestos and the incidence of mesothelioma. International journal of occupational and environmental health, 10(1), 22-25.

United Nations Commodity Trade Statistics Database. (2014). Statistics Division. Retrieved from http://comtrade.un.org/db/ce/ceSnapshot.aspx?px=HS&cc=2524 (Accessed July 2015).

Urban Poverty in India: Slums. (2012). Research for social & economic development. Poverties.org. Retrieved from http://www.poverties.org/urban-poverty-in-india.html (Accessed July 2015).

Van der Perk, M. (2007). Soil and Water Contamination: From Molecular to Catchment Scale. CRC Press.

Van Zandwijk, N.; Clarke, C.; Henderson, D.; Musk, a. W.; Fong, K.; Nowak, A.; and Penman, A. (2013). Guidelines for the diagnosis and treatment of malignant pleural mesothelioma. Journal of Thoracic Disease (Vol. 5).

Virta, R. L. (2001). Some Facts About Asbestos. Factsheet FS -012-01, (March), 1–4. Retrieved from http://www.capcoa.org/Docs/noa/%5B12%5D%20USGS%20Facts%20on%20Asbestos.pdf (Accessed July 2015).

Virta, R. L. (2002). U. S. Department of the interior U. S. Geological survey Asbestos : Geology, Mineralogy, Mining, and Uses, 1–28.

Virta, R. L. (2005). Mineral Commodity Profiles Asbestos. U.S. Geological Survey Circular. Retrieved from http://pubs.usgs.gov/circ/2005/1255/kk/ (Accessed August 2015).

Wachowski, L.; and Domka, L. (2000).Sources and Effects of Asbestos and other Mineral Fibers Present in Ambient Air. Polish Journal of Env. Studies; 9; 443-454.

Waste Management Company Munich (Abfallwirtschaftsbetrieb München). http://www.awm-muenchen.de/wir-ueber-uns/auftrag.html

WHO. (2006). WHO Workshop on Mechanisms of Fiber Carcinogenesis and Assessment of Chrysotile Asbestos Substitutes, November 8–12, 2005, Lyon, France Retrieved from http://www.who.int/ipcs/publications/new_issues/summary_report.pdf (Accessed August 2014).

WHO. (2012). National Programmes for Elimination of Asbestos-Related Diseases: Review and Assessment. World Health Organization.

WHO. (2014a). Asbestos: elimination of asbestos-related diseases. World Health Organization.

WHO. (2014b). Chrysotile Asbestos. World Health Organization.

Wisconsin Department of Natural Resources (2007). Asbestos – History and Uses Retrieved from https://web.archive.org/web/20071228042633/http://www.dnr.state.wi.us/org/aw/air/reg/asbestos/asbes3.htm (Accessed May 2015).

Woodson, R. D. (2012). Construction Hazardous Materials Compliance Guide: Asbestos Detection, Abatement and Inspection Procedures. Elsevier.

World Bank Group. (2009). Good Practice Note: Asbestos: Occupational and Community Health Issues. Operations Policy and Country Services.

Wylie, A. G.; Bailey, K. F.; Kelse, J. W.; and Lee, R. J. (1993). The importance of width in asbestos fiber carcinogenicity and its implications for public policy. American Industrial Hygiene Association Journal, 54(5), 239-252.

Zbyszek, O.; and Wladek, M. (1997). Processing of X-Ray Diffraction Data Collected in Oscillation Mode. Methods in Enzymology. Macromolecular Crystallography. 276. 307-326.

10 APPENDICES

Appendix 1: Fiber length quantification from Single Fibers of MY_1-2-3_SE and MY_1-1-4_SE SEM image

Fibersample no.	F_Length(in μm)
1	5.799264
2	5.910432
3	10.703008
4	4.40426
5	11.10522
6	12.89626
7	3.432312
8	3.48172
9	7.845332
10	4.895252
11	5.353048
12	2.630204
13	13.804904
14	3.667772
15	3.730304
16	14.870264
17	2.189392
18	2.799272
19	2.851768
20	11.450304

Appendix 2: Fiber width quantification from Single Fibers of samples MY1-2-3SE and MY1-1-4_SE

Fiber sample no.	F_width (in µm)
1	0.121976
2	0.094184
3	0.10036
4	0.10422
5	0.079516
6	0.083376
7	0.098816
8	0.10036
9	0.098816
10	0.060216
11	0.088008
12	0.090324
13	0.093412
14	0.09264
15	0.080288
16	0.10422
17	0.104992
18	0.086464
19	0.095728
20	0.06562

Appendix 3: SEM Image of MY_1-2-3_SE

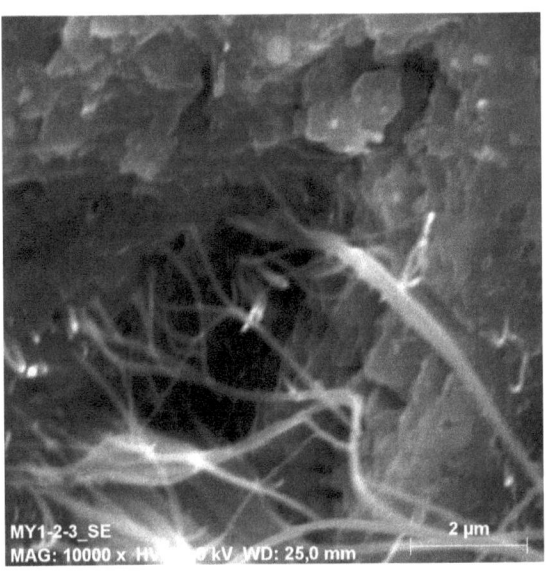

SEM Image MY_1-1-4_SE

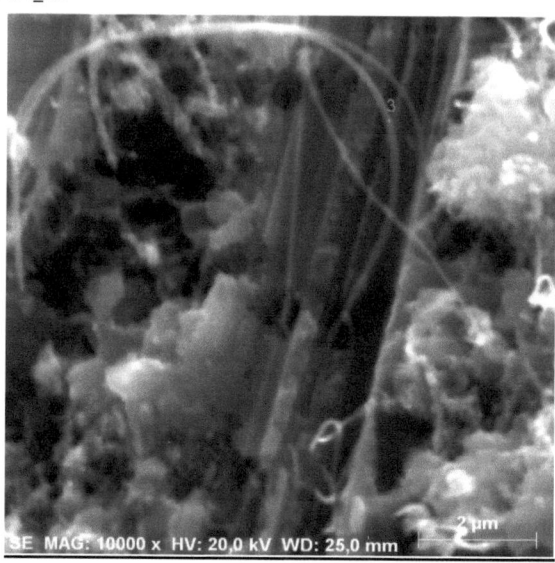

Appendix 4: Standards specified under Factory Act and Mines Act

Subject matter relating to recommended code of practices	BIS Code
Method for determination of airborne asbestos fibre concentration in work environment by light microscopy	IS: 11450:2006
Recommendations for safety and health requirements relating to occupational exposure to asbestos	IS: 11451-1986 (Reaffirmed 2005)
Recommendations for control of emission of asbestos dust in premises manufacturing products containing asbestos (Asbestos cement products)	IS: 11770 (Part 1) - 1987
Recommendations for control of emission of asbestos dust in premises manufacturing products containing asbestos (Friction materials)	IS: 11770 (Part 2) - 2006
Recommendations for control of emission of asbestos dust in premises manufacturing products containing asbestos (Non-cement asbestos products other than friction materials)	IS: 11770 (Part 3) (Reaffirmed 2001) - 1987

Recommendations for cleaning premises and plants using asbestos fibres	IS: 11767 - 2005
Recommendations for disposal of asbestos waste material	IS: 11768 - 1986 (Reaffirmed 2005)
Guidelines for safe use of products containing asbestos (Asbestos cement products)	IS: 11769 (Part1) - 1987
Guidelines for safe use of products containing asbestos (Friction materials)	IS: 11769 (Part2) - 1986
Guidelines for safe use of products containing asbestos (Non-cement asbestos products other than friction materials)	IS: 11769 (Part3) - 1986
Recommendations for personal protection of workers engaged in handling asbestos	IS: 12078 - 1987 (Reaffirmed 1997)
Recommendations for packaging, transport and storage of asbestos	IS: 12079-1987 (Reaffirmed 1997)
Recommendations for local exhaust ventilation systems in premises manufacturing products containing asbestos	IS: 12080 – 1987 (Reaffirmed 2001)
Recommendations for pictorial warning signs and precautionary notices for asbestos and products containing asbestos (Workplaces)	IS: 12081 (Part-1) - 1987
Recommendations for pictorial warning signs and precautionary notices for asbestos and products containing asbestos (Asbestos and its products)	IS: 12081 (Part-2) - 1987
Recommendations for the selection, use and maintenance of respiratory protective devices	9263 – 1980

Appendix 5: Properties of various types of Asbestos

Types/Properties	Chemical Formula	Color	Essential Composition	Structure
Chrysotile	$Mg_3Si_2O_5(OH)_4$	White, Grey, Green	Mg silicate with some water	Usually highly fibrous fibers, fine and easily separable
Crocidolite	$NaFe_3^{2+}Fe_2^{3+}Si_8O_{22}(OH)_2$	Lavender Blue, Metallic Blue	Na, Fe silicate with some water	Fibrous in iron stones
Amosite	$Fe_7Si_8O_{22}(OH)_2$	Ash Grey, Brown	Fe, Mg silicate with some water	Lamellar or coarse to fine fibrous and asbestiform
Tremolite	$Ca_2Mg_5Si_8O_{22}(OH)_2$	Grey white, Greenish, Yellowish, Bluish	Ca, Mg silicate with some water	Long or prismatic and fibrous aggregates
Anthophyllite	$(Mg, Fe)_7Si_8O_{22}(OH)_2$	Greyish white, Brown grey or Green	Mg silicate with some water	Lamellar or fibrous asbestiform
Actinolite	$Ca_2(Mg, Fe)_5Si_8O_{22}(OH)_2$	Greenish	Ca, Mg, Fe silicate with some water	Reticulated long prismatic crystals and fibers

[Ca: calcium; Fe: iron; Mg: magnesium; Na: sodium; Si: Silicon; O: Oxygen; OH: Hydroxide. Virta, 2005]